THE ROLE OF
TRACE METALS IN PETROLEUM

THE ROLE OF
TRACE METALS IN PETROLEUM

T. F. Yen

Associate Professor of Chemical Engineering,
Environmental Engineering Sciences,
and Medicine (Biochemistry)
University of Southern California, Los Angeles

ann arbor science PUBLISHERS INC.

P.O. BOX 1425 ● ANN ARBOR, MICHIGAN 48106

PREFACE

Petroleum has been known since the dawn of modern civilization. Although large quantities have been recovered and consumed, the basic science of petroleum is still developing. Specifically, very little is known about the role of metals in petroleum. Although the amount of metals present in petroleum is trace, the impact of successful industrial processing on clean environmental control is immense. Trace metals present in petroleum are important both in the genesis of petroleum and its refining. Furthermore, the consequence of emission and erosion is directly related to metals. An understanding of this nature will increase the knowledge of the scientists as well as of the energy-concerned public.

This book is a general survey of the nature of trace metals in petroleum, followed by the analytical methods for determination of metals in petroleum. The topics of the metals present in by-rpoducts of petroleum are then discussed. Basic information on geochemistry, bondings, demetallation and newly discovered metals such as molybdenum is presented. Finally, the important problem of recovery of these trace metals for resource use is raised.

This book developed from the symposium on "The Role of Trace Metals in Petroleum" sponsored by the American Chemical Society, Chicago, August 1973, of which the present editor was the chairman. Through this period, the zeal and interest of many authors never diminished, and for this I am grateful. I also thank Mrs. Darlene Baxter and many other students for their help.

T. F. Yen
May, 1975

CONTENTS

CHAPTER 1

CHEMICAL ASPECTS OF METALS IN NATIVE PETROLEUM

T. F. Yen
Departments of Chemical Engineering
Environmental Engineering, and Medicine (Biochemistry)
University of Southern California
Los Angeles, California 90007

The following is a status report of the metals
present in native petroleum. First, the occurrence
of the metals is discussed with two important topics--
metalloporphyrins and nonporphyrin metals--presen-
ted. At present, no review of nonporphyrin is
available. Highlights of this review center on the
topics of biogenesis of porphyrins and also vanadium
and nickel. Finally a few remarks are made on the
applications. The objective of this chapter is to
illustrate the fact that as basic knowledge of these
metals is acquired, intelligent use and a novel,
practical method will be developed for the processing as well as exploration of petroleum. Trace
metal in petroleum is the link between its formation
from basins and its refinement into the final products. Understanding the true nature of the role of
trace metals in petroleum will help the progress of
petroleum industries.

Native petroleum falls into the class of raw
fuel, which itself is not "clean." Aside from
"clean" hydrocarbons, petroleum contains the following contaminants:

(a) nonhydrocarbons, which are heterocyclics containing
    sulfur, nitrogen, and oxygen
(b) minerals such as silica and metals
(c) high molecular weight, large asphaltic molecules.

The concentrations of these contaminants may be
minor, but they are the source of environmental pollutants and the cause of corrosion of equipment and
poisoning of processing catalysts. The refining

process serves to eliminate these contaminants by
chemical conversion and upgrading. Contaminant (a)
is eliminated by reductive cleavage of heterocycles
by splitting of heteroatoms, S, N, O into H_2S, NH_3,
and H_2O. For contaminant (b) the metals can be
removed by a slurry process using asphaltenes.
Contaminant (c) is converted using the usual depoly-
merization of asphaltic stacks by means of hydrode-
sulfurization. Therefore, to ensure a clean end
fuel from native petroleum, desulfurization, denitrification and demetallization are necessary. The
removal of metals is a more complex problem due to
the following:

1. Metals are chelated or complexed in ligands that
    are completely compatible in petroleum and make
    separation difficult.
2. The amount of metals is very small, usually ranging
    from 1-10,000 ppm; so far vanadium is the highest.
3. Metals are generally associated with contaminant
    (a), the heterocycles, and contaminant (c), the
    asphaltic fraction.
4. Metals can exhibit catalytic effects during conver-
    sion. Demetallization may cause odd effects.
5. Refining and upgrading must involve the use of
    catalysts, which can be poisoned by metals.

Metals in petroleum, then, can be important in
the sense that they hold the key to industrial processing. The role of metal in petroleum also helps
in exploration, prospecting and oil pollution abatement.

OCCURRENCE

All native petroleums contain some inorganic
constituents. A spectrographic determination of 23
domestic crude oils reveals 28 elements from ash.[2]
These metals are:

U, Zn, Zr, V, Sr, Sn, Pb, Ni, Nd, Mo, La, Ga, Cu, Ca,
Cr, Co, Ce, Ba, B, As, Ag, K, Na, Mg, Mn, Tl, Fe, Al

The more abundant metals are V and Ni. Table 1.1
lists the metal content of V, Ni, Cu and U in 23
domestic oils together with wet ash analysis of
these oils.
As a rule of thumb, both vanadium and nickel
increase with the asphaltic content of the crude oil
(API gravity is an indicator). The lighter oils
contain less metal. With some exceptions, the

Table 1.1

Metal Contents of Typical United States Petroleum[2]

Location State and Field	Stratigraphic Unit	Age	% Ash	% of Ash			
				V	Ni	Cu	U
Arkansas							
Schuler	Jones sand	Jurassic	0.041	3.7	2.5	0.35	0.0006
Stevens-Smart	Hesston fm.	Cretaceous	0.004	4.4	5.4	0.84	0.0012
Colorado							
Gramps	Dakota ss.	Cretaceous	0.001	4.2	7.6	0.26	0.0017
Gramps	Morrison fm.	Jurassic	0.001	9.6	16.	0.26	0.0060
Kansas							
Brewster	Arbuckle gr.	Ordovician	0.002	13.	8.4	1.6	0.0014
Coffeyville	Upper part of Arbuckle gr.	Ordovician	0.002	16.	5.0	0.22	0.0012
Iola	Bartlesville sand	Pennsylvanian	0.130	1.2	0.69	0.01	0.0002
Montana							
Big Wall	Heath sh.	Mississippian	0.012	20.	11.	0.10	0.0001
New Mexico							
Table Mesa	Dakota ss.	Cretaceous	0.003	0.01	0.09	5.6	0.0018
Oklahoma							
Kendrick	Fort Scott ls.	Pennsylvanian	0.001	8.4	2.2	13.	0.0032
Lafoon	Wilcox sand of drillers	Ordovician	0.022	20.	9.2	0.08	0.0009
Utah							
Roosevelt	Green River fm.	Tertiary	0.004	0.41	7.6	0.13	0.0001

Table 1.1, continued

Location State and Field	Stratigraphic Unit	Age	% Ash	% of Ash V	% of Ash Ni	% of Ash Cu	% of Ash U
Wyoming							
Elk Basin	Tensleep ss.	Pennsylvanian	0.010	38.	9.2	0.16	0.0003
Grass Creek	Curtis fm.	Triassic	0.038	28.	6.4	0.10	0.0004
Halfmoon	Embar fm.	Pennsylvanian and Triassic	0.058	17.	4.8	0.26	0.0003
Halfmoon	Tensleep ss.	Pennsylvanian	0.023	22.	0.10	0.10	0.0003
Hamilton Dome	Curtis fm.	Triassic	0.038	28.	6.4	0.10	0.0004
Hamilton Dome	Embar fm.	Pennsylvanian and Triassic	0.012	46.	7.6	0.16	0.0003
Hamilton Dome	Madison ls.	Mississippian	0.038	28.	7.0	0.07	0.0003
Lost Soldier	Wall Creek ss. mbr.	Cretaceous	0.001	5.6	7.2	0.26	0.0024
Oregon Basin, N	Embar fm.	Pennsylvanian and Triassic	0.015	40.	7.6	0.03	0.0043
Oregon Basin, N	Madison ls.	Mississippian	0.035	22.	6.4	0.03	0.0022
Oregon Basin, N	Tensleep ss.	Pennsylvanian	0.018	40.	8.2	0.22	0.0075

vanadium content is higher than the nickel content (Table 1.2). The ratio V/Ni has been used as a parameter of age but there is no definite correlation.

Table 1.2

Metal Contents of Some Heavy Crudes (ppm)

No.	Crude Oil	°API gravity	V	Ni
SR	Boscan	10.3	1200	100
KS	Heavy Mara	17.5	961	85.7
QM	Laqunillas	16.3	455	58.3
SL	Taparito	---	393	42
CY	Bachaquero	19.0	355	54.5
DW	Hamaca	6.0	330	83.3
DV	La Canoa	15.2	322	93.5
SP	Cabimas	22.0	280	32
SS	Merey	17.0	205	47
KT	Heavy Orinoco	18.6	199	50.8
SQ	Barinas	26.5	158	71
SM	Leona	24.1	126	36
SO	Temblador	20.0	48	43
DL	Baxterville	---	33	15
OR	Kuwait	---	16	4.4
PL	Ragusa	19.9	8.5	68.6
---	West Texas*	---	4.8	3.3
---	San Joaquin*		2.8	0.95
---	Ordovician*		.09	.23

*Actually it is light oil

Currently, the metallic components in petroleum can be classified into the following categories:

   (a) Metalloporphyrin* chelates
      (aa) vanadyl and nickel porphyrins
      (ab) chlorophyll a and other hydroporphins*
      (ac) highly aromatic porphin chelates
      (ad) porphyrin decomposition ligands
   (b) Transition metal complexes of tetradentate mixed ligands such as V, Ni, Fe, Cu, Co and Cr.

*Porphyrin used here signifies the alkyl substituted analogs derived from porphin, which is the parent skeleton.

(ba)   simple complexes* from resin molecules
(bb)   chelates* from asphaltene sheets
(c)   Organometallic compounds such as Hg, Sb and As.
(d)   Carboxylic acid salts of the polar functional
        groups of resins, such as Mo and Ge.
(e)   Colloidal minerals, such as silica and NaCl.

There are only two major classes of metals of
primary importance in the above list--the porphyrin
metals and the nonporphyrin metals.   The porphyrin
metals belonging to (aa) have been studied widely.
The nonporphyrin metals have not yet been explored.
The nonporphyrin metals could be (ab), (ac), and
(ad) because in these categories the porphin skeleton has lost the physical properties of a typical
porphyrin due to either interrupted conjugation or
increased aromaticity.   The more important types of
nonporphyrin metals are those of the (b) category.
In summary, the following definition is applied:

Porphyrins   (aa)
Nonporphyrins   (ba)
                  (bb)
                  (ab)--only to very little con-
                         tributions
                  (ac)
                  (ad)

In the following section we will concentrate
more on exploring the present status of metal complexes of porphyrins and nonporphyrins, whereas for
categories (c), (d), and (e) only a passing remark
is made.   For (c) both alkyl and aryl can be possible, although the sandwich type of π-complexes for
arenes is also possible.   For (d) categories, the
germanium and molybdenum in coals can be complexed
with o-dihydroxyl groups of catechol.   There are
also other possibilities, such as thiocarboxylic
acids and other nitrogen ligands:

---

*Complexes here refers to the condition when the ligand mole-
cule to a single metal atom is plural, whereas for chelates
the ligand molecule can only be one.

Finally the (e) category is always carried over by formation waters and other fluids during recovery. The size usually is submicron and is difficult to remove.

METALLOPORPHYRINS

Just 40 years ago, Treibs[3] found that metalloporphyrins exist in bitumens, coals and shales. In these petroporphyrins he identified vanadium as well as iron. He established the well-known chemical association between chlorophyll in marine plants and the process of petroleum formation. It was not until 1948 that the second major metallic components in nickel were established.[4] In the intervening years, a number of investigators have found that petroporphyrins are concentrated in the gas oil, resin and asphaltene fractions.[5-8] Baker, *et al.* determined the yields of petroporphyrins from a number of asphaltenes of native crudes and other bituminous materials.[9] Data of ten sources of crude oil as well as tar sands, oil shale and gilsonite are listed in Table 1.3.

Table 1.3

Yields and Sources of Petroporphyrins

Asphaltene or Bitumen	Location	Age	Yields ppm
Agha Jari	Iran	L. Miocene	120
Baxterville	Mississippi	M. Cretaceous	28
N. Belridge	California	Pleistocene-Pliocene	3100
Boscan	Zulia, Venezuela	Eocene-L. Oligocene	1800
Burgan	Kuwait	M. Cretaceous	220
Mara	Zulia, Venezuela	M. Cretaceous	300
Melones	Anzoategui, Venezuela	U. Oligocene	17
Rozel Point	Utah	Recent	17
Santiago	California	U. Miocene	38
Wilmington	California	Miocene	570
Athabasca tar sands	Alberta	L. Cretaceous	145
Green River oil shale	Colorado	Eocene	55
Gilsonite	Utah	L. Eocene	40

From the study by means of mass spectrometry, electron spin resonance (ESR), and electronic and fluorescence spectrometry in recent years the following facts are known:
(a)   The extractable fossil porphyrins are in general inhomogeneous. They are identified as a series of alkyl homologs of cycloalkanoporphin. The first series is identified as the etioporphyrin III (Etio) and the second series is identified as deoxophylloerythroetioporphyrin (DPEP) (Figure 1). The alkyl substituents of both series are in the range of 27 to 41. The actual substitution sites are not known. Both Etio and DPEP series complexed with V and Ni are found.[9]

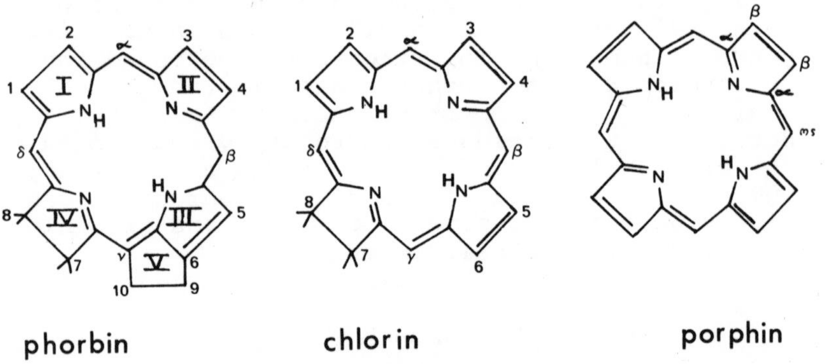

phorbin   chlorin   porphin

Figure 1.1.   Nomenclature of porphyrin system. Chlorin is the basic skeleton of natural chlorophylls. Notice the 7,8-dihydro group and pentacyclic ring V. Chlorin does not contain ring V. Porphin is the basic skeleton of porphyrins (alkyl porphins). Etioporphins contain methyl and ethyl groups in the β-positions, etioporphyrin I is M, E, M, E, M, E, M, E (4-fold symmetry); etioporphyrin II is M, E, E, M, M, E, E, M (2-fold symmetry); etioporphyrin III is M, E, M, E, M, E, E, M (no symmetry) and etioporphyrin IV is M, E, E, M, E, M, M, E (plane symmetry) along the β-sites of 1, 2, 3, 4, 5, 6, 7, 8 positions.

(b)   The extent of these distributions and their shapes and ratios are found to be a geological parameter for estimating the age of a given petroleum. Samples with wide distribution are more matured. Samples with narrow distribution bands are of non-marine origin. As the age of petroleum increases, the petroporphyrin proceeds from DPEP to Etio (Figure 2).[10]

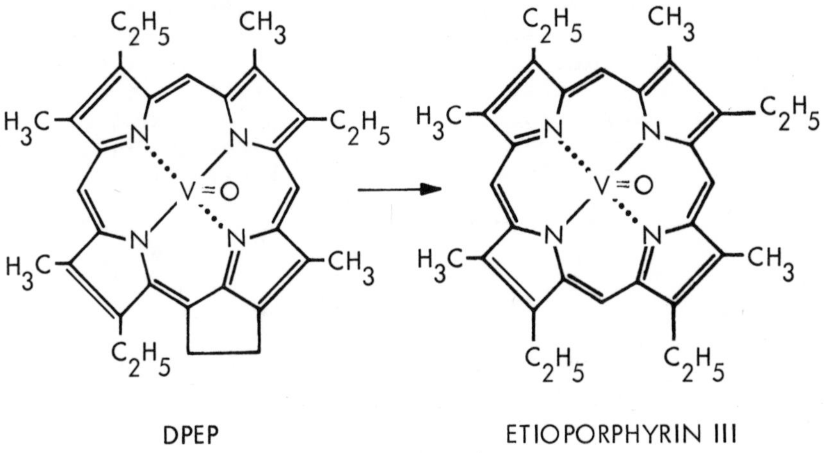

DPEP   ETIOPORPHYRIN III

Figure 1.2. The DPEP type and etio type porphyrins. Substituents on bridge and on pyrrole ring are
possible. The alkyl groups may vary.

(c) Monobenzoporphin and cycloalkanomonobenzo-
porphin series as in (b) also have been found (Figure 1.3). Exact mass measurement indicates that the
the rhodo type spectra (visible) exhibited by these
benzoporphins do not contain carbonyl (oxygen)
groups. This is a minor series,[11] which usually
consists of samples of old age. Thomas and Blumer[13]
also reported these series for a Triassic sediment.

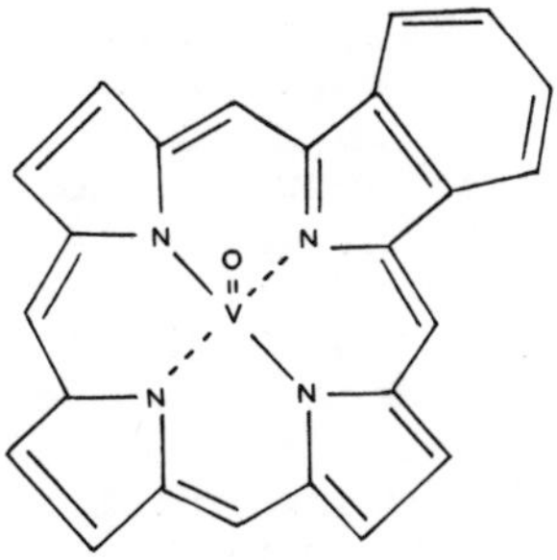

Figure 1.3. A benzoporphin structure. The benzo group could
fuse either on ring I, II, III, and IV or any
other combinations.

(d)   Another minor series of porphin was found in which the same porphin nucleus bears two isocyclic C-5 rings. This skeleton is unknown in nature.[11] In order to account for this, it is possible that more than one bridge-substitution can occur in chlorine[6], e.g., 2 and α. The vinyl group at 2 of pheophobin as can undergo hydration and condensation at α-bridge (see Figure 1.4).

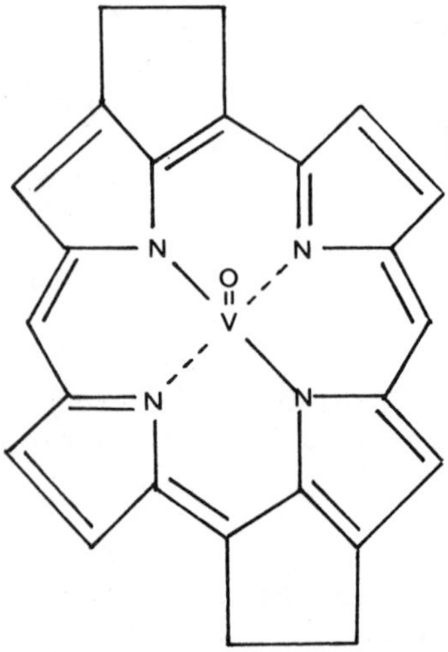

Figure 1.4.   Porphin structure with two cycloalkano rings.
Besides ring V there is another ring linking
the 2 and α positions.

(e)   Chlorophyll a is detected in sediments from the Dead Sea.[12] This may be a very limited and unique case. Usually chlorophylls will demetallize near the sediment-water interface. For example pheophytin a was found in marine sediments[27] and dihydropheophytin a in deep sea core.[15]

(f)   Phylloerythrin, which is with carbonyl functional groups, is present in recent sediments such as Rozel Point tar of Utah.[14] Other intermediates are also present: the intact ester has been reported, e.g., methylpheophobide from Eocene coal.[16]

(g)   Chlorins and mesochlorins are found in a number of deep ocean sediments.[15] For the DPEP and Etio type of porphins isolated, only nickel porphyrins were found.

(h)   For Precambrian Nonesuch shale the porphins found were highly aryl type.[17] They include the benzo, tolyl and α-naphyl substituents.

Porphyrin-degraded products such as open-chain bile pigments were also detected. The stability of van-adyl m-tetra-α-naphyl porphin and other high aromatic derivatives is unusually high.[18]
From the above experimental facts, a general-ized theory concerning the diagenesis of petropor-phyrins is formulated and will be discussed in the section on biogenesis.

NONPORPHYRIN METALS

In the present definition, the nonporphyrins will also include the altered or modified porphin structure, such as hydroporphins (ab), arylporphins (ac), and porphin-degraded products (ad). All three classes have been included in the discussion of the above section since they are the secondary or ter-tiary derivatives or the precursors of the regular porphin structure (aa). In many cases they lose their porphyrin identity (properties) even when they exist in the same environment.
Evidences of the presence of highly aromatic-containing porphyrins in petroleum are as follows:

1. Direct parent mass and parent mass minus aryl group were observed for Nonesuch shale by mass spectrom-etry.[17]

2. The isotropic nuclear-spin coupling constant a_O is lower than those of typical porphyrins (delocal-ized). An increase of anisotropic coupling constant A_{11} is another measurement.[19]

3. Nitrogen superhyperfine splitting in asphaltic fractions was observed.[20]

4. Difficulty was encountered in demetallation by using strong acid for petroleum asphaltics.[11]

5. A green porphyrin fraction, which exhibited 590 nm in the visible spectrum, has been isolated from petroleum. This fraction has been demonstrated to be a ring-containing porphyrin.[9]

The other major class of nonporphyrin metals is the complexes of the tetradentates of mixed ligands. These ligands can be any combination of four atoms from N, S, and O. They can be small, the (aa) class, or large, the (ab) class. The for-mer class is formed from polar resin molecules and the latter is formed from asphaltic sheets. The ligand sites could be as follows:

4N*, porphyrins
3N10
2N20*, β-ketimines
1N30, N(2-hydroxyphenyl)salicylidenimine
40, β-diketones
301S*
202S, monothio-β-diketones
103S
4S* dithiocarbonates
3S1N*
2S2N*
1S3N
S02N
SN20
N02S

Of the above combinations, the ones with an aster-
isk are already found in resin molecules, $e.g.$, in
a Baxterville resin fraction by mass spectrometry.[21]
A number of 2N20 model complexes of vanadium have
been synthesized,[52-54] correlations of these complexes
have been made, and a method for identification of
the mixed ligand type has been developed.[55-57] The
following additional facts indicate that there exist
quadridentate-mixed ligands in petroleum.

1. A fraction of the vanadium complex from
Sdom crude does not give Soret absorption peaks,
and is easily demetallated.[22] Most oxovanadium(IV)
complexes exhibit d-d transitions ($d_{xy} \rightarrow d_{x^2-y^2}$ (transition II) and $d_{xy} \rightarrow d_{z^2}$ (transition III) at much
lower (10^3 less) extinction coefficients when compared with that of the Soret peak ($a_{1u} \rightarrow e_g*$). Vanadium complexes devoid of the Soret peak would suggest
there is no extreme conjugation. In the case of
mixed donor tetradentates, demetallation proceeds
with ease. Therefore it is plausible that there
are tetradentates in Sdom crude.

2. Certain fractions of Boscan vanadium petroporphyrins only give even masses by mass spectrometry.[23] The only possibility is that these fractions
contain odd numbers of nitrogen atoms in the donor
system. Furthermore, Boscan and Baxterville resin
fractions exhibit intense parent masses at odd
numbers.

3. Magnetic susceptibility measurement on a
Boscan (SR) vanadium petroporphyrin extract indicates Curie-Weiss dependence.[11] V-V interaction or

V-O-V super exchange may indicate a monomeric vanadium complex of one or three nitrogen atoms.

   4. Asphaltenes can have a fixed capacity for
complexing metals.[24] This fact is demonstrated
with washing cycles and with increments of scattering by small angle scattering experiments.[25] These
sites may be defect centers from the aromatization
of S, N, and O-containing hydrocarbon skeletons.[26]

BIOGENESIS

   Petroporphyrins originate from chlorophylls.
There are two questions that arise immediately:
How is the magnesium in chlorophylls removed and
how are vanadium and nickel introduced?

$$MP + 2H^+ \rightleftarrows PH_2 + M^{2+}$$

MP here is metalloporphyrin and PH_2 is free base
porphyrin. In general, the forward equation is demetallation and the reverse is metallation. The
dissociation of metalloporphyrins is governed by
chemical stability.

water	Ag(I), K(I), Na(I), Pb(II, Hg(II)
dil. HCl	Zn(II), Cd(II), Mg(II), Fe(II)
conc. H_2SO_4	Cu(II), Mn(II), Co(II), Ni(II),
	Fe(III)
HBr-acetic acid	VO(IV)
trichloroacetic acid	VO(IV)*

From the above data, it is obvious that pheophytinization (removal of magnesium) is easy, whereas
devanadylation (removal of VO) is difficult. It is
also foreseeable that pheophytinization can be accomplished in natural conditions, perhaps soon after
decaying and possibly before burying.
   In supporting this, chlorophyll a is only observed in limited cases, yet pheophytin a as well
as other intermediates such as pheoporphyrin and
phylloerythrin have been identified in recent sediments. This scheme may be represented in Figure 1.5
where chlorophyll a is gradually changed to deoxophylloerythrin.[14] Again the origin of porphyrin in
sediments is chlorophyll pigment only, such as chlorophyll a, chlorophyll b, *Chlorobium* chlorophyllys

---

*Porphyrin is tetra-ms-4-biphenylporphin.

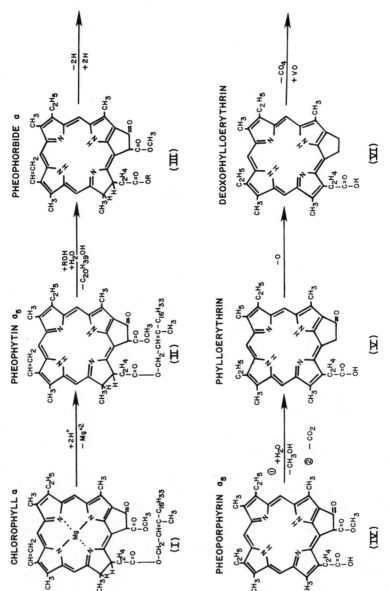

Figure 1.5a.   Diagenic scheme from chlorophyll a to deoxophylloerythrin and then to DPEP.

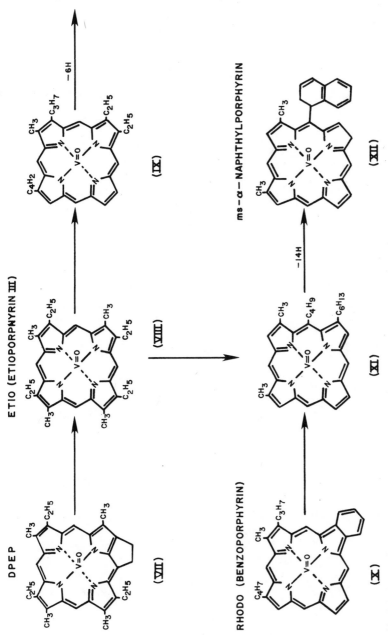

Figure 1.5b. Hypothetical scheme from DPEP to benzoporphyrin and ms-α-naphthyl porphyrin.

650 and 660 (Figure 1.6).[9] Some contributions of
pigments from chlorobacteiaceae are possible since
the alkyl substituents in petroporphyrin usually
exceed 34 carbons (usual change for petroporphyrin
is from 27-41). The upper limit for chlorophyll *a*
is 34 carbons excluding the phytyl group.

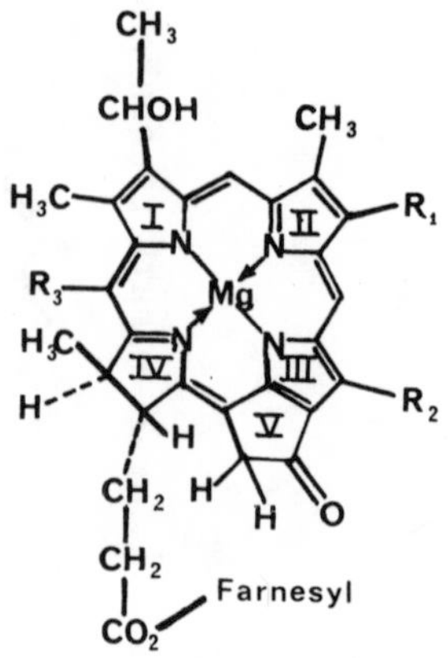

650 Series	R_1	R_2	R_3
Fraction 1	isobutyl	ethyl	H
2	n-propyl	ethyl	H
3	isobutyl	methyl	H
4	ethyl	ethyl	H
5	n-propyl	methyl	H
6	ethyl	methyl	H
660 Series	R_1	R_2	R_3
Fraction 1	isobutyl	ethyl	ethyl
2	isobutyl	ethyl	methyl
3	n-propyl	ethyl	ethyl
4	n-propyl	ethyl	methyl
5	ethyl	ethyl	methyl
6	ethyl	methyl	methyl

*Figure 1.6.   Structure of Chlorobium chlorophyll 650 and 660
based on Holt.*

The isolation of chlorin P_6 and mesochlorin P_6 and possibly the purpurin 18 intermediates[15] suggests that allomerization is an important process for chlorin formation by oxidative ring opening (Figure 1.7). It is also possible that petroporphyrins of the etio and phyllo type can be derived from these chlorins.

Largely the marine depositional environment is reducing; therefore, one has to consider the reducing sites of chlorophyll a, namely the 2-vinyl, 9-keto and the phytyl double bond. Under reducing environments, removal of phytyl groups can be accomplished by elimination rather than by hydrolysis.[15] Both phytol and neophytadiene have been isolated. In the sequence outlined in Figure 1.8, under conditions of geological time spans, the 2-vinyl, 9-keto will undergo reduction. Products of phytyl double bond reduction have already been reported.[15] Furthermore, epimerization of 10-position is possible. These stepwise structural changes do have significant geochemical implications and could be further explored.

Under drastic geological conditions, the decarboxylation of $7-CH_2-CH_2COOH$ and the ring scission of cyclopentane ring can be realized, *i.e.*, the conversion to DPEP. The conditions can only be realized when the biostratinomy and fossilization processes are at depths where geothermal temperatures are elevated (150°C). If age can be traded with temperature,[28] certainly this conversion of decarboxylation has already taken place in recent sediments.[15]

The next important questions are when and why the vanadium and nickel are introduced. Answering the second part is easier; it has already been found that the segregation of metal in freshwater shale is different from those of marine shale.[29] For example, in freshwater shale, the metals are Cu, Pb, Zn and Sn, whereas in marine shale the metals are Ni and V. This concentration of Ni and V in marine shale has been interpreted to be due to the accumulation of marine organisms. Selective metals in the aquatic organisms of both brown algae and marine animals are enriched from 10-280,000 times (Table 1.4). The case of V and Ni should be noticed in particular. For example, in ascidia and holothurians, vanadium pigments are found in blood cells or vanadocytes. Vanadium content in various forms of ascidia was found to contain from 1-6520 ppm on a dry basis.[30] Also in *Phaleusia manillata*

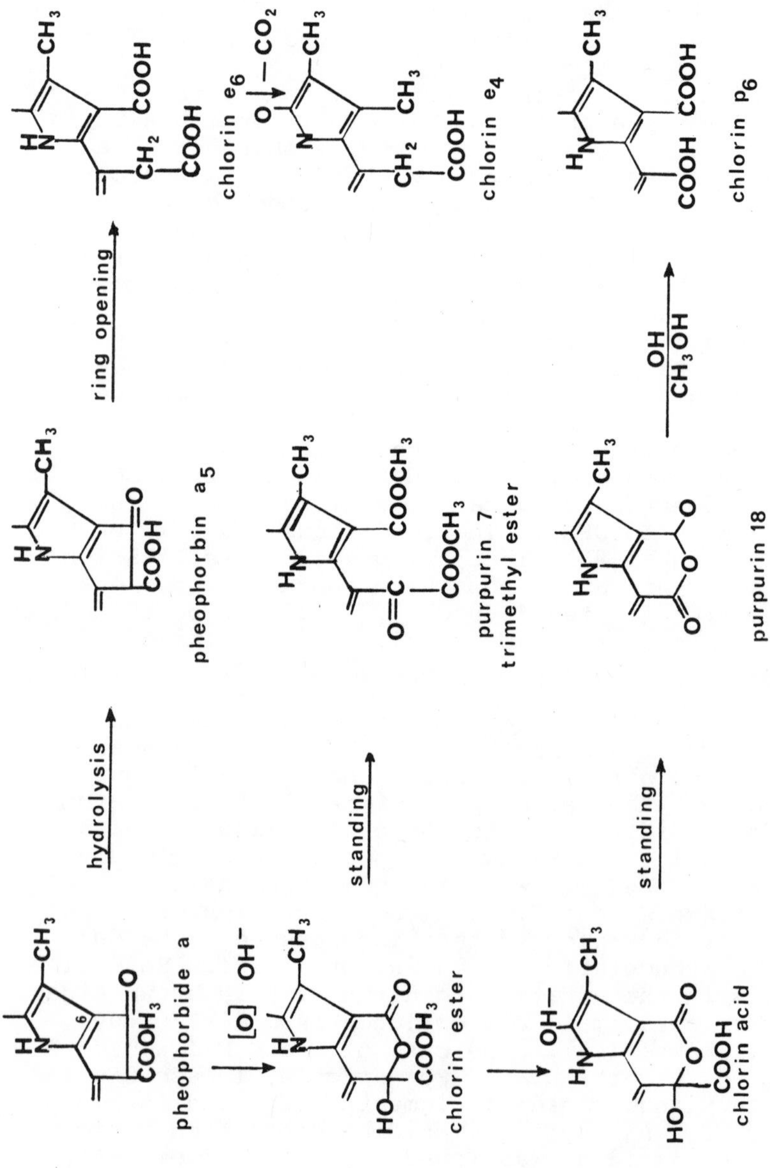

Figure 1.7.   Allomerization scheme and the formation of chlorin.

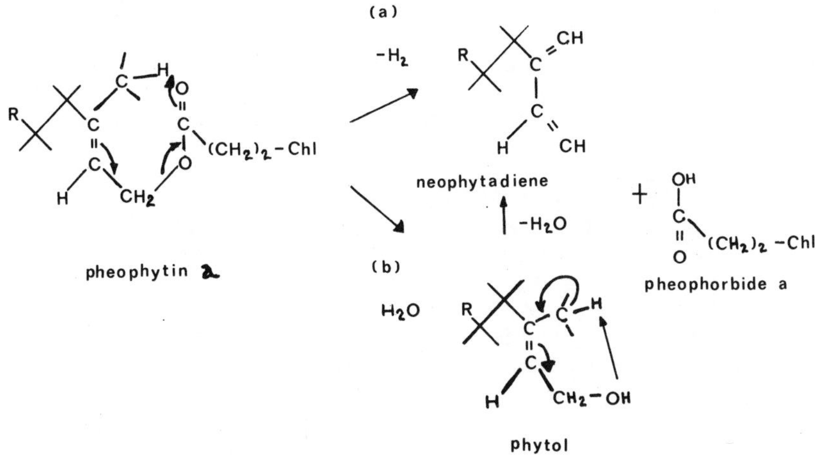

Figure 1.8.   Pheophytimization scheme.

Table 1.4
Enrichment Factors of Marine Organisms

	Brown Algae (Fresh Weight)		Marine Animals (Dry Weight)	
	I^a	II^b	I	II
Ni	200-1000	1600-1800	5,000	41,000
V	10-300	>160-5000	17,000	>280,000
Zn	400-1400	400-1400	32,500	32,500
Co	(>4500)	(13,000)	>7,000	21,000
Mn	17,000c	---		
Fe	10,700c	---		
Cu	330-880c	---		

[a]Data from W. A. P. Black and R. C. Mitchell (Ref. 60, excluding c.
[b]Data from I. Noddack and W. Noddack (Ref. 61).
[c]Data from E. D. Goldberg (Ref. 62). The values are the average for *Fucus sorratus* and *F. vesiculosus*.

80% of total V is in blood, whereas in *Ascidia mentula* only 10% is in blood.[31] Clams, scallops and oysters are rich in NI; for example, *Cordium edule* (common edible cockle) contains 20 ppm on a dry basis.[33] Peleypoda have been found in some beds in the Carboniferous era.

At present it is without doubt that petroleums originate from marine organisms.[33]   The introduction of nickel should be easier (under less drastic conditions) than that of vanadium, due to the relative stability.   Recent sediments are richer in nickel than the ancient sediments.   This certainly is valid when comparing the stability of the metalloporphyrins. At this stage it is more likely that the vanadylation and nickelation of porphyrins occurs during the diagenic process.   It is unlikely that a process will be introduced via formation water by secondary migration.   The subsurface water during migration contains no appreciable amount of vanadium and nickel, regardless if the source of water is vadose, connate, or juvenile.   On the other hand, traces of alkali and alkaline earth metals may be derived from contacting these waters through the migration process.

The next important transformation is the scheme of DPEP to Etio.   Again, this ring opening is due to the geothermal temperature that can be illustrated by depth.[10]   A good illustration is the ratio of DPEP to Etio for a number of petroporphyrins as a function of depth of the petroleum (Figure 1.9). A question arises: What is the possibility of a 6-vinyl group formation during the ring scission to Etio?   The vinyl group intermediate caused by thermal cleavage could then hydrogenate to the ethyl group in the reducing environment.   This possibility will point to the existence of two cyclo-pentane rings to the porphin nucleus.   Evidently the 2-vinyl group could link to a-bridge.

Another important feature is transalkylation--the random rearrangement from etio or other alkyl porphins to a mixture of alkyl porphins based on a normal distribution pattern.[9,37]   Such a case has been tested in the laboratory; $e.g.$, vanadyl etio-porphyrin I pyrolyzed at 420°C will yield all other combinations of alkylporphins with substituents from one C to nine C as indicated by mass spectrometry (Figure 1.10).   One such possibility is from structures VIII to IX in Figure 1.10. Transalkylation is analogous to Jacobson's rearrangement of aromatic hydrocarbons and occurs only under high temperature history.[28]

Finally, another point must be made here for the aryl-substituted or fused porphin.   These can occur only if the environment for dehydrogenation is possible. Dehydrogenation can occur only under extremely high temperatures (or through geological

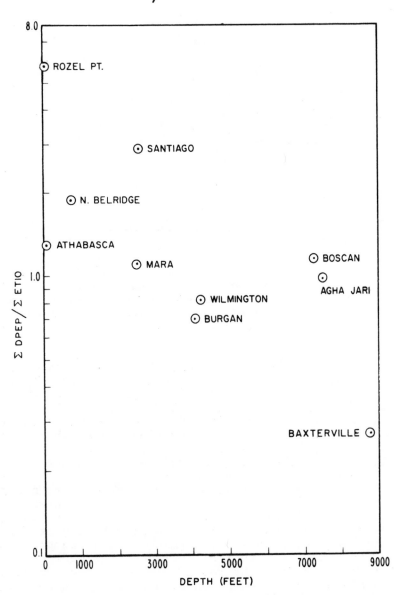

Figure 1.9.   A plot of the ratios of DPEP to Etioporphyrins
versus the depth of burial of a number of native
petroleums from which these porphyrins derived.

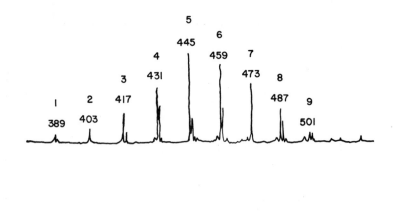

Figure 1.10.   Mass spectra of pyrolyzed vanadyl etioporphrin I
              indicating transalkylation. The products con-
              tain 1 to 9 methylene groups in a distribution
              pattern of which the 5-methylene one is the
              richest.

eons if time can trade temperature). Precambrian
shale and ancient sediments contain such porphins.
It should also be noted that dehydrogenation of
tetra-ms-9-anthryl porphin would yield highly aromatic porphins such as of 24-ring system illustrated
in Figure 1.11.[58]   These porphins may exhibit graph-
ite-like properties such as electric conduction.
Actually, the vanadium associated with natural
graphite may be of this nature.

     Both resin and asphaltene fractions are minor
components of the total crude, yet they contain the
bulk (10-50%) of the nitrogen and sulfur contents
of the crude (Table 1.5).   The probability of more
than one resin molecule forming a metal complex is
quite high.   A number of tetradentates made of
mixed donor atoms of resin molecules (intermolecular
complexes) are present in Figure 1.12.   These complexes can be readily dissociated by dilute acid.

     The distribution of sulfur oxygen and nitrogen
in asphaltene is quite high.   During the genesis of
asphaltene the graphization process is interrupted

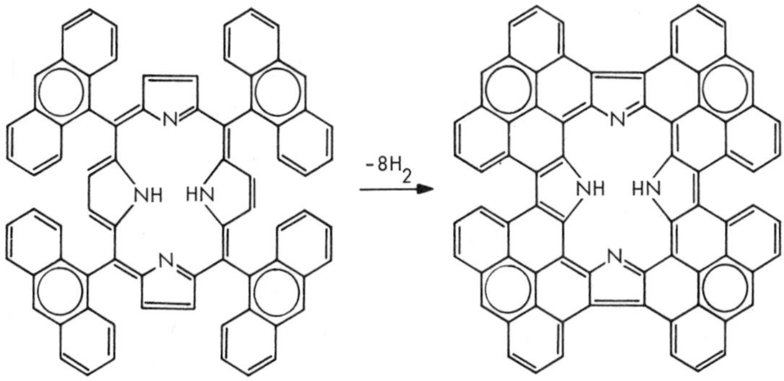

Figure 1.11.   Formation of highly-fused aromatic porphyrin system under high temperature.

due to the divalent nature of S and O and the tri-valent nature of N.   In this instance, the incomplete formation of tessellations of hexagons will prevent the complete aromatization of fused benzene network.   In this case "gaps" or "holes" will be formed and these centers are expected to be the sites of donor atoms (Figure 1.13).   Metals will be able to complex into these sites (Figure 1.14) and cannot be removed by aqueous acid washings.[24]

Porphyrin is known to undergo degradation in the presence of air or light.   Although this is less likely in native petroleum, it still remains as a possibility.   Photooxidation as well as photoreduction occurs readily in the scission of ring, which leads to open-chain analogs such as propendyopent-like derivatives.   β-Radiations are known to cause the loss of a-methane carbon to form bile pigments (Figure 1.15).

APPLICATIONS

As discussed above, there are a number of parameters from the study of the biogenesis of porphyrin that can yield important insights in exploration and prospecting.   With some of these parameters, *e.g.*, the ratio of DPEP to Etio, the bandwidth already is useful in approximating age in organic geochemistry.[10]   Other parameters such as V/Ni ratio have been used for correlation, although

Table 1.5

Distribution of Percentage of Total Nitrogen of Crude
in Resins and Asphaltenes

Crudes	%Resin	%Asphaltenes	%N in Resins	%N in Asphaltenes	%S in Resins	%S in Asphaltenes
Mara-La Lune	9.1	4.1	30.0	27.5	20.5	10.5
Oficina	3.9	1.1	18.3	12.8	33.3	13.8
Ragusa	9.2	0.28	—	48.2	—	24.3
Wilmington	14.2	5.1	30.5	15.6	22.9	8.5
North Belridge	18.0	5.1	25.8	14.8	23.9	7.8
Boscan	29.4	18.0	20.4	41.0	35.1	23.0
Sandhills	4.4	0.44	8.5	3.2	9.1	3.1
Abell-Ellenburger	4.2	0.24	39.3	3.1	33.8	3.9
South Waddell	3.9	0.39	50.0	10.0	17.4	3.1
Keystone	2.2	0.22	40.0	10.0	15.1	0.34
South Ward	1.2	0.80	14.4	3.9	6.7	2.3
Hiseville	0.97	0.19	53.0	4.1	12.2	0.7
Athabasca	24.2	19.4	21.8	49.7	28.2	29.5

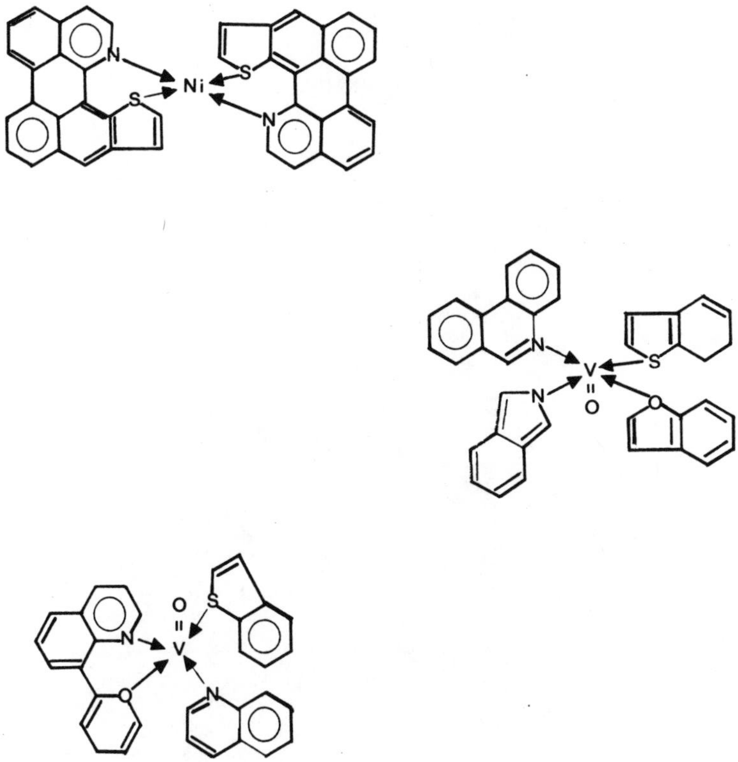

Figure 1.12. Metal complexes (tetradentates) of mixed donor atoms.

there is a conflicting trend. A profile analysis including all trace metal for a "fingerprint" is very helpful, *e.g.*, in identification for oil-spill.[34]

Since sulfur is known to bond as a donor atom to vanadium[35] as well as to nickel,[36] it should be anticipated that dehydrosulfurization proceeds proportional to metal removal.[38] It has been demonstrated that in reality this turns out to be so.[39,40] This is an important aspect in the processing of heavy oils, but a detailed discussion is not in the scope of this paper.

Metals in charge stocks are known to poison catalysts for further refining and cracking.[41,42] Methods are developed for removal of these metals.

Figure 1.13.   Molecular model indicating the metal uptake
          capacity of asphaltenes. The left upper gap is
          made of sulfur, nitrogen, and oxygen atoms.

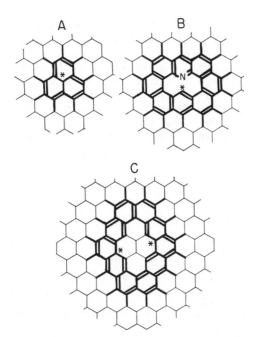

Figure 1.14.   Gap and hole defects of hexagonal network.

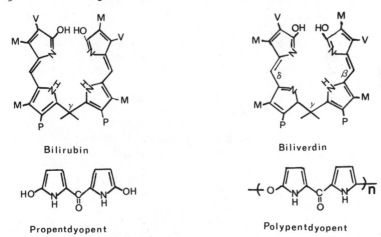

Bilirubin

Biliverdin

Propentdyopent

Polypentdyopent

Figure 1.15.   Degradation products of porphyrins.

In brief, ion exchange,[43] electron discharge,[44] and dialysis[45] appeared to have little success. The best method is still acid treatment. Groenning's procedure using glacial acetic acid-HBr is a traditional method.[46] Methanesulfonic acid,[48,49]

p-toluenesulfonic acid,[47] sulfuric acid,[50] and phosphoric acid[59] can be used for the demetallation of metalloporphyrins.

Finally, Edler[51] suggests the following: In order to eliminate the difficulties encountered during processing of high vanadium-containing petroleum, the use of a vanadium-containing cracking catalyst for heavy fractions or distillation residues must be sought.   Such contacts would avoid the poisoning suffered by conventional cracking catalysts.

ACKNOWLEDGMENT

Partial support of this work by PRF No. 6272-AC2 is acknowledged.

REFERENCES

1. Yen, T. F.   *Energy Sources*, *1(1)*, 117 (1973).
2. Ball, J. S., W. J. Wenger, H. J. Hyden, C. A. Horr and A. T. Myers.   *ACS Petroleum Chem. Div. Preprints*, *1* 241 (1956).
3. Treibs, A.   *Ann. der chem.*, *510*, 42 (1934).
4. Glebovskaya, E. A. and M. V. Vol'kenshtein.   *J. Gen. Chem.* (USSR), *18*, 1440 (1948).
5. Skinner, D. A.   *Ind. Eng. Chem.*, *44*, 1159 (1952).
6. Corwin, A. H., W. S. Caughey, A. M. Leone, J. E. Danieley and J. F. Bagli.   *ACS Div. Petroleum Chem. Preprints*, *2*, A-35 (1957).
7. Dunning, H. N., J. W. Moore, H. Bieber, and R. B. Williams.   *ACS Div. Petroleum Chem. Preprints*, *5*, 169 (1960).
8. Eldib, C. A., H. N. Dunning and R. J. Bieber.   *ACS Div. Petroleum Chem. Preprints*, *5*, 31 (1960).
9. Baker, E. W., T. F. Yen, J. P. Dickie, R. E. Rhodes, and L. F. Clark.   *J. Am. Chem. Soc.*, *89*, 3631 (1967).
10. Yen, T. F. and S. R. Silverman.   *ACS Div. Petroleum Chem. Preprints*, *14(3)*, E32-E39 (1969).
11. Yen, T. F., L. J. Boucher, J. P. Dickie, E. C. Tynan, and G. V. Vaughan.   *Proc. Inst. Petroleum*, *55*, 87 (1969).
12. Nissenbaum, A., M. J. Baedecker and I. R. Kaplan.   *Geochim. Cosmochim. Acta*, *36*, 709 (1972).
13. Thomas, D. W. and M. Blumer.   *Geochim. Cosmochim. Acta*, *28*, 1147 (1964).
14. Yen, T. F. in *Trace Substance in Environmental Health*, Vol. VI, D. D. Hemphill, Ed.   (Columbia: Univ. of Missouri Press, 1973) pp. 347-353.

15. Baker, E. W. and C. D. Smith. *ACS Div. Petroleum Chem. Preprints, 19(4),* 744 (1974).
16. Dilcher, D. C., R. J. Pavlick and J. Mitchell. *Science 168,* 1447 (1970).
17. Rho, J. H., A. J. Bauman, H. G. Boettger and T. F. Yen. *Space Life Sci., 4,* 69 (1973).
18. Vaughan, G. B., E. C. Tynan and T. F. Yen. *Chem. Geol., 6,* 203 (1970).
19. Yen, T. F., E. C. Tynan, G. B. Vaughan and C. J. Boucher, in *Spectrometry of Fuels* (New York: Plenum Press, 1970) pp. 187-201.
20. Yen, T. F., *Naturwissenschaften, 58,* 267 (1971).
21. Dickie, J. P., R. L. Anderson, J. R. Boal and T. F. Yen. 16th Annual Conference of Mass Spectrometry, Pittsburgh, Pa., May 1968.
22. Branthaver, J. F., API Project 60 Advisory Meeting, Laramie, Wyoming, July 1967.
23. Sugihara, J. M. and G. Y. Wu.   Private communication.
24. Erdman, J. G. and P. H. Harju. *J. Chem. Eng. Data, 8,* 252 (1963).
25. Pollack, S. S. and T. F. Yen. *Anal. Chem., 42,* 623 (1970).
26. Yen, T. F., J. G. Erdman and A. J. Saraceno. *Anal. Chem., 34,* 694 (1962).
27. Orr, W. L., K. O. Emery and J. R. Grady. *Bull. Am. Assoc. Petrol. Geologists, 42,* 925 (1958).
28. Yen, T. F., in *Chemistry in Space Research,* R. F. Landel and A. Rembaum, Eds.   (New York: Elsevier, 1972) pp. 105-153.
29. Degens, E. T., E. G. Williams and M. L. Keith. *Bull. Am. Assoc. Petrol. Geol., 41,* 2427 (1957).
30. Bertrand, D. *Bull. Am. Museum Natur. Hist., 94(7),* 407 (1950).
31. Nason, A. in *Trace Elements,* C. A. Lamb, O. G. Bentley and J. M. Beattie, Eds.   (New York: Academic Press, 1958) p. 269.
32. Silverman, S. R. *Proc. 8th World Petroleum Congress, Vol. I* (New York: Elsevier, 1971) pp. 47-54.
33. Bertrand, G. and R. Paulais. *Compt. Rend., 203,* 683 (1936).
34. Filby, R. H. and K. R. Shah. *Proc. Am. Nucl. Soc. Conf. on Nucl. Methods in Env. Res.* (Columbia: Univ. of Mo. Press, 1971) p. 86.
35. Sugihara, J. M., T. Okada and J. F. Branthaver. *J. Chem. Eng. Data, 10,* 190 (1965).
36. Thompson, M. C. and D. H. Busch. *J. Am. Chem. Soc., 86,* 3651 (1965).
37. Baker, E. W. *J. Am. Chem. Soc., 88,* 2311 (1966).
38. Larson, O. A. and H. Beuther.   Private communication.

39. Ebel, R. H.   *ACS Div. Petroleum Chem. Preprints, 17(3),* C46 (1972).
40. Drushel, H. V.   *AÇS Div. Petroleum Chem. Preprints, 17(4)* F92 (1972).
41. Mills, G. A.   *Ind. Eng. Chem., 42,* 182 (1950).
42. Mills, G. A. and H. H. Shabaker. *Pet. Refiner, 30,* 97 (1951).
43. McClintock, T. L.   MS Thesis, Rensselaer Polytechnic Institute, June 1950.
44. Harju, P. H. and T. J. Hardwick.   Private communication, 1961.
45. Wolsky, A. A. and F. W. Chapman, Jr.   Midyear Meeting API Div. Refinery, May 1960.
46. Groenning, S.   *Anal. Chem., 25,* 938 (1953).
47. Kotova, A. V., S. V. Emelyanova and V. G. Ben'kovskii. *Khim. Tekhnol. Topliv. i Masel, 10,* 29 (1965).
48. Rho, J. H., A. J. Bauman, T. F. Yen and J. Bonner. *Science 167,* 754 (1970).
49. Erdman, J. G.   U. S. Pat. 3,190,829 (1965).
50. Dean, R. A. and R. B. Girdler.   *Chem. Ind., 100* (1960).
51. Edler, E.   Abstract Paper D, Section IV, International Symposium on Vanadium and Other Metals in Petroleum, Universidad del Zulia, Maracaibo, Venezuela, Aug., 1973.
52. Boucher, L. J., E. C. Tynan and T. F. Yen. *Inorg. Chem., 7,* 731 (1968).
53. Boucher, L. J. and T. F. Yen. *Inorg. Chem., 7,* 2665 (1968).
54. Boucher, L. J. and T. F. Yen. *Inorg. Chem., 8,* 689 (1969).
55. Boucher, L. J., E. C. Tynan and T. F. Yen, in *Electron Spin Resonance of Metal Complexes,* T. F. Yen, Ed. (New York: Plenum, 1969) pp. 111-130.
56. Yen, T. F.   Papers presented at Gordon Research Conferences on Geochemistry, Plymouth, New Hampshire, Aug. 1970.
57. Dickson, F. E., C. J. Kunesh, E. L. McGinnis and L. Petrakis. *ACS Div. Petroleum Chem. Preprints, 16(1),* A37 (1971).
58. Yen, T. F.   Papers delivered at Joint Conference on Possible Organic Superconductors, Jet Propulsion Laboratory, Pasadena, Calif., March 1970.
59. Baker, E. W. in *Organic Geochemistry,* G. Eglinton and M. T. J. Murphy, Eds. (Berlin: Springer-Verlag, 1969) pp. 464-497.
60. Black, W. A. P. and R. L. Michell.   *J. Marine Biol. Assoc., 30,* 575 (1952).
61. Noddack, I. and W. Noddack. *Arkiv. zool.,32A(4)* (1939).
62. Goldberg, E. E. in *Chemical Oceanography,* Vol. I, J. P. Riely and G. Skirrow, Eds.   (New York: Academic Press, 1965) pp. 163-196.
63. Holt, A. S. in *The Chlorophylls,* L. P. Vernon and G. R. Seely, Eds. (New York: Academic Press, 1966) pp. 111-118.

CHAPTER 2

THE NATURE OF METALS IN PETROLEUM

R. H. Filby
Department of Chemistry and Nuclear Radiation Center
Washington State University
Pullman, Washington

Although petroleum consists predominantly of hydrocarbons, most petroleums contain measureable quantities of many metals. Nickel and vanadium are commonly the most abundant metals but Fe, Zn, Cr, Cu, Mn, Co, and others are almost always present in concentrations ranging from less than 1 ng/g to more than 100 μg/g. The nature of these metals and their abundances in crude oils can give information on the origin, migration, and maturation of petroleum as well as providing a basis for regional geochemical prospecting. Also, the question of at what stage and how metals were incorporated during petroleum genesis is an intriguing geochemical problem, one that is far from being solved. The nature of metals in crude and residual oils is also of interest to the refinery operator and to environmentalists concerned with emissions from oil-fired power plants.

Several early studies reported the presence of trace elements in the ashes of oils[1,2] but the first quantitative data were presented by Shirey[3] who determined 13 metallic elements in the ashes of 7 oils. Most of the work published since World War II has concerned the geochemical significance of trace element distributions within oil fields and related rocks or the nature of Ni and V in petroleum as these elements cause problems in refining processes.

Bonham[4] analyzed 60 U.S. oils for V, Cu and Ni by a spectrographic technique, and Erickson, Myers and Horr[5] analyzed crude-oil ashes for V, Ni, Cu, As, Co, Mo, Pb, Cr, U and Mn by a semiquantitative procedure. In both studies, the authors concluded that the elements detected occurred in an oil-soluble form. Bonham[4] showed that oils from certain basins contain distinctive trace element suites but that correlations of producing strata from pool to pool could not be made.

Hodgson[6] in a study of the oils of Western Canada measured Ni, V, and Fe and concluded that the V/Ni ratio decreased with increasing maturation as indicated by the sequence of asphaltic shallow oils (Cretaceous) to deeper less asphaltic oils (Devonian). Hodgson[6] interpreted the data to indicate that the V porphyrin complex was less stable than the Ni complex and that that the contents of Ni, V, and Fe were derived from the original source organic material. Hyden[7] and Ball, et al.,[8] however, have presented data to show that the V/Ni ratio increased with age of the host rock. Similar conclusions were used by Al-Shahristani and Al-Atyia[9] to show vertical migration from deeper Cretaceous formations to shallow Tertiary formations in Iraqi fields. Other authors[10,11] have concluded that the V/Ni ratio does not correlate with age.

In recent years several studies of the distribution of trace elements in U.S.S.R. petroleums and correlations with geochemical properties have been made.[12-16] Kotova, et al.[12] showed that V, Ni, Cu, Ge and Ga varied as a function of depth and age of reservoir rock and that the metals were concentrated in the asphaltic fraction of the oils. Similar conclusions were reached by Katchenkov and Flegentova,[13] Botneva,[14] Mileshina, et al.[15] and Nurev and Dzhabharova[16] but Gilmanshin, et al.[11] concluded that V, Ni, Fe and Cu were not related to the asphaltene or resin contents of oils of the Pashiisk region.

For valid geochemical interpretation of trace element data in crude oils it is necessary to know in what forms the trace elements occur, and except for Ni and V such knowledge is largely lacking. The metals may be present in oils as inorganic particulate matter (such as mineral grains or absorbed on clay minerals), in emulsified formation waters, introduced as drilling fluids or corrosion inhibitors or present as organometallic complexes. Only if the metals are present in an oil-soluble form as true complexes can meaningful geochemical information be obtained from trace element data.

Nickel and vanadium in crude oils have been studied extensively because of the presence of Ni and V porphyrins, which are thought to have been derived from chlorophyll (or hemoglobin), thus indicating a biogenic origin for petroleum. It is now recognized[17,

that significant amounts of Ni and V, which cannot
be accounted for by free metalloporphyrins, are present in crude oils. The nonporphyrin Ni and V is
present in the asphaltic component of crude oils and
either occupies sites bounded by hetero-atoms (N, S, O)
or is present in metalloporphyrins strongly associated
by π-π bonding to the asphaltene aromatic sheets.[19]
The origin of this nonporphyrin Ni and V is not known.
It may have been incorporated in the asphaltene structure from the original organic source material, replaced other metal cations in the asphaltenes, or
incorporated by complexation from aqueous or solid
phases during the migration or maturation of petroleum.
Sugihara, et al.[18] have postulated that the nonporphyrin V in asphaltenes was the source of V introduced
into the porphyrin structure during the conversion
of chlorophyll in the source material to the metalloporphyrins found in oil. The fact that V and Ni may
occur in porphyrin and non-porphyrin forms may explain
some of the conflicting data concerning the correlations of V/Ni ratios with age of reservoir rock.

Very little is known of the nature of metals
other than Ni and V in crude oils. In several studies
cited previously[4],[5],[8],[12] it was demonstrated that the
metals occurred in oil soluble form, and some authors
have noted the association of metals with the asphaltic
fraction of petroleum.[12-16] Erdman and Harju[19] have
thrown doubt on these conclusions and consider that,
except for Ni and V, the trace element contents of
crude oils can be reduced to insignificant levels by
centrifuging or other procedures designed to remove
suspended particulate matter or a dispersed aqueous
phase.

The paucity of information on metals, other than
Ni and V, in crude oil is due principally to the poor
sensitivity of most common analytical techniques for
trace metals in oil. Neutron activation analysis has
been used for trace element measurement in crude oils
by several workers[20-24] and allows detection of many
elements at the sub-ppm level in small samples
(0.1-1.0 g). Shah, et al.[22],[23] have developed an instrumental neutron activation analysis method for the
determination of 23 elements in crude oils and Hitchon,
et al.[25] have used this method for a detailed study
of the trace element geochemistry of Alberta crude
oils. A review of the application of neutron activation analysis to trace metals in petroleum is included
in this book.[26]

The preliminary study reported here is part of
an investigation of the abundances and nature of trace-
element species in crude oils. Neutron activation

analysis was used to determine whether the trace
elements of a California Tertiary oil were present
in oil-soluble form or in an associated aqueous or
solid phase. The California oil (C-1) selected
was previously shown to have anomalously high concentrations of several trace elements, including
Ni, Cr, Co, Zn, Fe, Hg, Cu and As. The distribution
of trace elements in the oil was determined by separating the oil into three components—a methanol
soluble fraction, a methanol insoluble-n-pentane
soluble fraction ("resin") and an n-pentane insoluble
fraction (asphaltenes). Each fraction was separated
into molecular-weight fractions by gel permeation
chromatography (GPC), and liquid chromatography on
Al_2O_3 or SiO_2 was used to separate porphyrin concen-
trates in an attempt to determine whether porphyrins,
other than those of Ni and V, were present in the oil.

EXPERIMENTAL

Separation of Oil Components

Several crude oils were examined for use in the
distribution study. It was necessary to choose an
oil with no aqueous phase or observable mineral matter dispersed in the organic matrix. In addition
the fact that small fractions isolated from the oil
were to be analyzed required that the oil selected
have a high asphaltene content (>5%) and a high content of trace metals. The oil chosen was a Tertiary
California oil which had been analyzed previously by
neutron activation analysis.[24]
The oil chosen for detailed analysis was too
viscous for direct centrifuging and was diluted 1:1
with benzene (Baker "Analyzed" Reagent). The oil
was then centrifuged for 2 hours and the supernate
filtered through a Whatman #42 filter paper. (In
later experiments a final filtration step through a
5 μ Millipore filter was used.) After removal of
benzene by vacuum evaporation, the oil (1-10 g) was
weighed into a 250-ml polyethylene bottle. 200-ml
n-pentane (Baker Reagent) was then added and the
bottle shaken for 4 hr at 25°C to precipitate the
asphaltenes. The solution was centrifuged and the
asphaltenes were washed several times with n-pentane
until the wash solution was clear. The asphaltenes
were then dried at 45°C to remove the solvent and
the precipitation with n-pentane repeated. The small
amount of asphaltenes precipitated was combined with

the bulk of the asphaltenes. The maltenes were
recovered by evaporation of the n-pentane and were
then extracted with 75 ml of anhydrous methanol
(Baker "Analyzed") for 8 hr. The solution was centrifuged and the precipitated resins extracted twice
with methanol. The methanol was evaporated to give
the methanol-soluble fraction. The resins, asphaltenes
and methanol-soluble fraction were weighed and the
proportions of the three fractions determined.

Gel Permeation Chromatography Separations

     Gel permeation chromatography (GPC) of the
petroleum fractions was carried out on Sephadex LH-
20 (Pharmacia Fine Chemicals), with an approximate
exclusion limit of molecular weight 1500, and Styragel
1000 (Waters Associates) with an approximate exclusion limit of molecular weight 32,000. A standard
GPC column (Sephadex SR 25/45) was filled with LH-
20, preswollen with benzene methanol (90:10 v/v);
the final column dimensions were 35 x 2.5 cm. The
interstitial volume of the column (30.1 ml) was
determined by the breakthrough of a high molecular
weight asphaltene fraction (MW > 8,000) and the
column was calibrated using mesoporphyrin dimethyl
ester and phaeophytin (a+b) as described by Blumer
and Snyder.[27] The Sephadex GPC separation was carried out using 100-250 mg of the asphaltene, and
the absorbance at 526 nm was used to monitor the
separation. Material of molecular weight greater
than 1500 was excluded from the resin.
     The Styragel 1000 column (1.0 x 42 cm) was pre-
pared from resin swollen with benzene-methanol
(90:10 v/v). Polystyrene standards were used for
molecular weight calibration, and an exclusion limit
of 28,000 was determined. The Sephadex-excluded
asphaltene fraction was chromatographed on the
Styragel 1000 column and a continuous elution curve
was obtained from molecular weight 22,000 to 1000.
The eluates from the Sephadex and Styragel experiments were combined to give four arbitrary molecular
weight fractions: (a) < 1000, (b) 1000-4000, (c) 4000-
8000 and (d) 8000-22000. The relative proportions
of the four fractions were determined by weighing
after evaporation of the solvent. Similar separations were carried out on the resins and the methanol-
soluble component.

Determination of Metalloporphyrin
Contents of Oil Fractions

The direct spectrophotometric determination of metalloporphyrins in separated oil fractions was carried out by measurement of the Soret peak area as proposed by Sugihara and Bean.[28]   Although the nature of the different Ni and V porphyrin species in the California oil was not known, most petroleum porphyrins are either deoxophylloerythroetioporphyrins or etioporphyrin complexes.[29]   Metalloporphyrin contents were calculated from Beer's law

$$A = \varepsilon\ b\ c$$

where

A = area of Soret peak (integrated absorbance) in nm

ε = weighted extinction coefficient of Soret peak in liter nm. mol.$^{-1}$ cm^{-1}

b = path length (1 cm)

c = concentration of metalloporphyrins (Ni + V) in mol l^{-1}

As both Ni and V porphyrins contribute to the Soret peak, ε was calculated from the data in Table 2.1. The average Ni/V ratio for the three oil components was 15.2, and this value was used to calculate a weighted extinction coefficient used for all metalloporphyrin determinations.

Table 2.1

Soret Extinction Coefficients for Ni Etioporphyrin I and VO Etioporphyrin I[28]

Porphyrin	Soret Maximum (nm)	ε	Ni/V
Ni Etioporphyrin I	396	3.35×10^6	--
VO Etioporphyrin I	410	4.71×10^6	--
Oil Fractions	398	3.44×10^6*	12.5

*Weighted extinction coefficient using Ni/V = 12.5

A Ni/V ratio of 6.25 gives a weighted extinction coefficient at 3.56×10^6 liter nm cm^{-1}mol^{-1}, which is only 3% different from that used in this study. Small variations in the Ni/V ratio with molecular weight do not materially change the calculated metalloporphyrin content. The spectrophotometric method was checked by activation analysis. Nickel and vanadium porphyrin aggregates separated by liquid chromatography porphyrins (LC) were dissolved in benzene and the solution analyzed both by the spectrophotometric technique and neutron activation analysis. The results are shown in Table 2.2. The determination of metalloporphyrins in the asphaltene and resin fractions of high molecular weight must be regarded as only approximate since the Soret area was often less than 10% of the total absorbance over the Soret wavelength region.

Table 2.2

*Comparison of Metalloporphyrin Contents
Measured by Spectrophotometry and NAA*

Metalloporphyrin	*Concentration* $\mu mol/g$	
	NAA	*Spectrophotometric*
Ni	4.2×10^{-3}	3.8×10^{-3}
VO	6.6×10^{-3}	7.2×10^{-3}

Liquid Chromatography

Selected oil fractions of low molecular weight were chromatographed on SiO$_2$ or Al$_2$O$_3$ by methods described by Hodgson, *et al.*[30] and Dunning and Rabon.[31] Elution with hexane-benzene (9:1) gave a fraction high in Ni porphyrins and elution with benzene gave a fraction enriched in V porphyrins. Chromatography on Al2O3 was used to purify these fractions.

Determination of Elemental Concentrations

The determination of Ni, V, Fe, Co, Ni, Cr, Cu, As, Sb, Hg and Zn was carried out by neutron activation analysis. The separated fractions and crude oils were analyzed by the methods of Shah, *et al.*[22],[23] and Filby and Shah.[26] To determine the contributions of the solvents used in separations to the trace element contents of the oil fractions, neutron activation analysis was used to measure the pentane, methanol and benzene trace element values.

RESULTS AND DISCUSSION

Table 2.3 shows the contents of Ni, V, Co, Fe, An, Cr, Sb, As and Na in the crude oil C-1 before and after centrifuging and filtering. Filtration of the oil produced a very fine black deposit on the filter, but analysis of the filter showed only background values of the trace elements measured. Also included in Table 2.3 are results of filtration and centrifugation of another California Tertiary oil (C-2) and Boscan crude oil of Venezuela. The table indicates that the oil C-1 shows no sifnificant differences in trace element content between the treated and untreated oil. The relative standard deviations of the results in Table 2.3 are approximately 5-10%. The oil C-2, however, shows significant decreases in As and Na, and the Boscan oil has lower values of Hg, Fe, As and Na in the filtered oil.

An aqueous phase (0.025 g/g) and a black solid phase (0.4 mg/g) were separated from oil C-2. The aqueous phase contained 2800 µg/g Na, 21.3 µg/g Fe and 18.2 µg/g As and appeared to be residual formation water. The solid phase contained 120 µg/g Na and 1496 µg/g Fe and some organic material. The aqueous phase contributed 0.34 µg/g As and 53.0 µg/g Na to the unfiltered crude oil As and Na concentrations, but negligible amounts of other elements could be ascribed to the aqueous phase. The solid phase contributed negligible amounts of all trace elements to the bulk crude oil. The results indicate that trace element data on crude oils should be interpreted with caution and that only centrifuged and filtered oils should be analyzed if geochemical information is desired.

The oil C-1 was also extracted for 4 hr with double-distilled water, and the extracted oil and water were analyzed to determine whether water-soluble metal compounds were present. The results,

Table 2.3

Elemental Concentrations in Filtered and Centrifuged Oils

Nature	Ni	V	Co	Hg	Fe	Zn	Cr	As	Sb	Na
					Elemental Concentrations ($\mu g/g$)					
California C-1										
Crude	93.5	7.5	12.7	21.2	73.1	9.32	0.634	0.656	0.0517	11.1
Filtered	98.4	7.7	13.5	23.1	68.9	9.76	0.640	0.655	0.056	13.2
California C-2										
Crude	113.0	6.0	13.9	1.49	77.2	19.5	0.685	1.63	0.061	65.2
Filtered	117.0	5.8	14.4	1.41	70.1	21.5	0.608	1.09	0.059	10.0
Boscan, Venezuela										
Crude	117.0	1120.0	0.198	0.139	16.9	0.619	0.380	1.20	0.273	25.0
Filtered	109.0	1110.0	0.178	0.0269	4.77	0.692	0.430	0.284	0.303	20.3

shown in Table 2.4, indicate that 92% of the Na,
37.5% of the As and 22.5% of the Sb in the oil were
extracted but that other elements were not significantly affected. Further extractions with water
did not change the Na, As or Sb concentrations of
the oil; thus either the three elements are partly
present in water-soluble form or centrifuging did
not remove an emulsified aqueous phase. Indirect
evidence against the latter conclusion is that
during precipitation of the asphaltenes with n-
pentane a solution of the maltenes was observed with
no visible evidence of an aqueous phase. The maltenes extracted from the n-pentane solution contained
8.80 µg/g Na (oil: 11.1 µg/g Na) and 0.504 µg/g As
(oil: 0.564 µg/g) and comprised 95% by weight of the
oil. If an aqueous phase had been present it should
have precipitated with the asphaltenes and more than
90% of the Na should have been removed from solution.
This was not the case; hence it may be concluded
that 92% of the Na in oil C-1 is present as an oil-
soluble compound that is also water soluble or is
hydrolyzed, perhaps as the Na salt of a petroleum
acid. Thus 37.5% of the As and 22.5% of the Sb in
the oil appears to be present as water soluble or
combined in hydrolyzable compounds.

Table 2.4

Extraction of Trace Elements from Oil C-1
by Distilled Water

	Concentration (µg/g)		
Element	Extracted Oil	Extractable	Crude Oil
Ni	92.0	1.0	93.5
Co	14.5	0.005	12.7
V	7.3	0.1	7.5
Fe	82.4	1.0	73.1
Zn	10.2	1.5	9.32
Cr	0.665	0.1	0.634
Hg	19.2	0.2	21.2
Sb	0.040	0.031	0.517
As	0.410	0.276	0.656
Cu	1.03	0.4	0.93
Se	0.391	0.1	0.364
Na	0.79	15.2	11.1

Table 2.5 shows the distribution of trace elements among the three components separated from
the crude oil, the relative proportions of each component, and the elemental ratios of asphaltene/crude
oil and resins/crude oil. The results show that the
asphaltenes contain the highest concentrations of
all elements and that the asphaltic component of the
oil (resins + asphaltenes) accounts for the major
part of the trace elements, V, Ni, Co, Fe, Hg, Zn,
Cr and Sb. Calculations show that 48% of the As
is present in the methanol-soluble fraction of the
oil, which suggests a low-molecular-weight compound,
possibly an alkyl or aryl arsine. The asphaltene/
crude elemental ratios show that Ni, V, Co, Fe, Cr,
Hg and Zn have similar fractional patterns. Antimony is strongly concentrated in the asphaltenes and
As shows the lowest enrichment of the elements studied.
The asphaltenes and resins exist in the oil in colloidal form,[32] and the crude oil system may be regarded as a transition from the polar aromatic micelle
of the asphaltenes to the less polar resins to the
nonpolar hydrocarbons of the bulk crude oil. The
trace elements concentrated in the asphaltenes may
be present in small highly polar molecules, which
would precipitate with the asphaltenes or might complex in the asphaltene sheet structure at sites
bounded by hetero atoms such as O, N, or S. Gel
permeation chromatography was used to investigate
this.
To determine the distribution of the trace
elements in the oil components as a function of
molecular weight, GPC was used on the methanol-
soluble component, resins and asphaltenes. The
weight distributions of the four arbitrary molecular-
weight fractions in the components are shown in
Table 2.6. The high-molecular-weight fraction of
the methanol-soluble component (>1500) was not further separated because of the small amount of sample
recovered. As expected, the relative amounts of the
four molecular-weight fractions are different for the
three components. The methanol-soluble fraction contains only 6.2% high-molecular-weight material whereas
the resins and asphaltenes contain 70.6% and 89.0%
high-molecular-weight material respectively. For
both resins and asphaltenes the 4000-8000 molecular-
weight fraction is the most abundant, and the heavi-
est molecular-weight fraction (8000-22000) was found
only in the asphaltenes (15.2%).
These findings are consistent with the nature
of the asphaltenes and resins as described by Yen[19]

Table 2.5

Distribution of Trace Elements in Components of Crude Oil C-1[a]

Concentration (μg/g)	Crude Oil	Methanol Soluble	Resins	Asphaltenes	$\frac{R}{C}$	$\frac{A}{C}$
% crude oil	100.0	57.5	37.5	4.99		
V	7.5	0.82	12.4	61.6	1.65	8.2
Ni	93.5	7.21	147.0	852.0	1.57	9.1
Co	12.7	0.8	10.7	122.0	0.84	9.6
Fe	73.1	1.95	66.4	895.0	0.91	12.2
Hg	21.2	0.686	29.6	140.0	1.40	6.6
Cr	0.634	0.300	0.894	7.540	1.41	11.9
Zn	9.32	0.74	8.86	109.	0.95	11.7
Sb	0.0517	0.0033	0.0130	1.22	0.25	23.6
As	0.656	0.546	0.290	2.25	0.44	3.4

a $\frac{R}{C}$ = ratio of concentration in resins to that in crude

$\frac{A}{C}$ = ratio of concentration in asphaltenes to that in crude

Table 2.6

Distribution of Molecular-Weight Fractions
of Oil Compounds

Molecular Weight Fraction	Percentage of Fraction		
	Methanol Soluble[a]	Resins	Asphaltenes
(1) 300-1,000	93.8	29.4	11.0
(2) 1,000-4,000	6.2	21.2	23.2
(3) 4,000-8,000	....	49.4	50.6
(4) 8,000-22,000	....	0	15.2
Total	100.0	100.0	100.0

a The higher-molecular-weight component of the methanol
   fraction was not separated into smaller fractions.

and others.[32] The asphaltenes contain 11% of low
molecular-weight material, which is consistent with
the idea that small polar molecules are associated
with the asphaltene micelles during precipitation but
are separated in the GPC procedure. The fact that
the resins and the asphaltenes contain material of
the same molecular weight range indicates the similar
nature of these materials, and the difference in precipitation behavior may be attributed to the higher
polarity of the asphaltenes as suggested by Witherspoon and Winneford.[32]
    The concentrations of nine trace elements in
each of the molecular-weight fractions of the resins
and the asphaltenes are shown in Tables 2.7 and 2.9
and the data for the methanol-soluble component are
given in Table 2.10. Vanadium was not determined in
the separated molecular-weight fractions because of
the small amount of material available for analysis.
For the asphaltenes, the highest trace-element concentrations, with the exception of Ni and Sb, are found
in the highest-molecular-weight fraction. The highest Ni was found in the 300-1000 fraction but this
is due entirely to Ni porphyrins, which are associated
with the asphaltenes. Antimony exhibits a behavior
contrary to that of all the other elements studied.

Table 2.7

Distributions of Trace Elements in Asphaltenes and Resins Fractions

Molecular Weight Fraction (GPC)	Asphaltenes or Resins (%)	Concentration (µg/g)								
		Ni	Co	Fe	Hg	Cr	Zn	Cu	Sb	As
Asphaltenes										
(1) 300–1,000	11.0	1327.0	2.67	480.0	72.0	0.77	112.00	0.34	11.0	0.850
(2) 1,000–4,000	23.2	189.0	30.00	368.0	20.9	4.80	52.00	1.50	0.9100	0.620
(3) 4,000–8,000	50.6	984.0	167.00	867.0	90.0	9.12	103.00	4.00	0.3500	1.900
(4) 8,000–22,000	15.2	1060.0	176.00	1934.0	350.0	19.6	225.00	7.20	0.1040	6.600
Total	100.0	852.0	122.00	895.0	140.0	7.540	109.00	3.02	1.2200	2.250
Resins										
(1) 300–1,000	29.4	206.0	4.37	30.1	22.0	0.310	3.31	<0.20	0.0430	0.407
(2) 1,000–4,000	21.2	110.0	10.00	24.0	44.0	0.800	11.00	<0.50	0.0026	0.200
(3) 4,000–8,000	49.4	80.2	24.90	236.0	72.0	2.960	27.00	1.30	0.0054	0.200
(4) 8,000–22,000	~0	--	--	--	--	--	--	--	--	--
Total	100.0	147.0	10.70	66.4	29.6	0.894	8.86	0.32	0.0130	0.290

Table 2.8

Distribution of Trace Elements in Asphaltenes*

Molecular Weight Fraction (GPC) Fraction (%)	300-1,000 11.0	1,000-4,000 23.2	4,000-8,000 50.6	8,000-22,000 15.2	Total 100.0
Element					
Ni	1327.0	189.0	984.0	1060.0	852.0
Co	2.67	30.00	167.00	176.00	122.00
Fe	480.0	368.0	867.0	1934.0	895.0
Hg	72.0	20.9	90.0	350.0	140.0
Cr	0.77	4.80	9.12	19.6	7.540
Zn	112.00	52.00	103.00	225.00	109.00
Cu	0.34	1.50	4.00	7.20	3.02
Sb	11.0	0.9100	0.3500	0.1040	1.2200
As	0.850	0.620	1.900	6.600	2.250

*All values in µg/g.

Table 2.9

Distribution of Trace Elements in Resins*

Molecular Weight (GPC) Fraction				
Fraction (%)	300–1,000	1,000–4,000	4,000–8,000	8,000–22,000	Total
Element	29.4	21.2	49.4	~0	100.0
Ni	206.0	110.0	80.2	—	147.0
Co	4.37	10.00	24.90	—	10.70
Fe	30.1	24.0	236.0	—	66.4
Hg	22.0	44.0	72.0	—	29.6
Cr	0.310	0.800	2.960	—	0.894
Zn	3.31	11.00	27.00	—	8.86
Cu	<0.20	<0.50	1.30	—	0.32
Sb	0.0430	0.0026	0.0054	—	0.0130
As	0.407	0.200	0.200	—	0.290

*All values in µg/g.

Table 2.10

Distribution of Trace Elements in Methanol Soluble Fractions

Molecular Weight Fraction	Total (%)	Concentration (μg/g)										
		Ni	Co	Fe	Hg	Cr	Zn	Cu	Sb	As
(1) 300-1,000	93.8	11.00	0.73	<1.00	0.410	--	<1.00	--	0.0046	0.340
(2) >1,600	6.2	<1.00	2.61	9.90	7.130	--	3.50	--	<0.0020	5.000
Total	100.0	7.21	0.80	1.95	0.886	<0.3	0.74	<0.5	0.0033	0.546

The highest Sb content is found in the lowest-molecular
weight fraction (11.0 µg/g) and the Sb concentration
decreases with increasing molecular weight of the
asphaltene fraction. This suggests that Sb is contained in small molecules of high polarity that are
insoluble in n-pentane and that precipitate with the
asphaltenes. The nature of the Sb compounds is un-
clear and further work is necessary. Possible compounds are the alkyl and aryl stibines R_xSbH_{3-x}.

Of the other elements in the asphaltene fractions,
Cr, Cu and Co show increasing concentrations with
increasing molecular weight of the fraction, indicating
that these elements are associated with the large
asphaltene molecules. The highest concentrations of
Fe, Hg, Zn, and As are found in the highest-molecular-
weight fraction of the asphaltenes but these elements
show higher concentrations in the 300-1000 molecular-
weight fraction than in the 1000-4000 fraction. For
Fe, Hg and Zn this may indicate the presence of porphyrin complexes or other organometallic compounds.
Iron porphyrins have been identified in shale oil,[33]
and Zn coproporphyrin occurs in animal feces.[34]
Neither of these metalloporphyrins has been found in
crude oil, however. Mercury forms porphyrin complexes
with mesoporphyrin[34] but the complexes are decomposed
by water. The extraction of crude oil C-1 with distilled water removed insignificant amounts of Hg, thus
indicating that Hg is not present in the oil as a
porphyrin complex. Arsenic also shows a higher concentration in the low-molecular-weight fraction than
in the 1000-4000 fraction. This is consistent with
the previous conclusion that part of the As in crude
oil is present in low-molecular-weight compounds such
as alkyl or aryl arsines.

The resin fractions show trace-element patterns
similar to those observed for the asphaltenes, although
the elemental concentrations in the resin fractions
are lower than in the corresponding asphaltene fractions.

The methanol-soluble component of the oil yielded
only 6.2% high-molecular-weight material. The trace-
element data show that only Ni and Sb have higher
concentrations in the low-molecular-weight fraction.
Nickel is present in this fraction entirely as Ni
porphyrins. Antimony is probably present in the low-
molecular-weight fraction as a low-molecular-weight
organoantimony compound as discussed previously.
The nature of the small amounts of Fe, Hg, Zn and Co
in the low-molecular-weight fraction is not known but
porphyrin complexes, although not identified in the
oil, cannot be ruled out.

The association of the typical porphyrin absorption spectra with high-molecular-weight compounds in crude oil [27,28] has led to the assumption that either the porphyrin conjugated system preserves its spectral identity even in such large molecules as the asphaltenes or that simple metalloporphyrin molecules are associated with the asphaltene sheet structures. Sugihara, *et al.*[18] have separated a number of Ni and V porphyrins (phyllo, etio, etc.) from Boscan asphaltenes and concluded that the metalloporphyrins in the high-molecular-weight fraction of the asphaltenes were associated with much larger molecules and thus appeared to be of much higher molecular weight in the GPC separations. Other evidence for the association of metalloporphyrins with high-molecular-weight compounds has been provided by Hodgson, *et al.*[35] A number of authors have pointed out that most crude oils contain insufficient porphyrins to account for the Ni and V content of the oil; thus these elements are also present in crude oils as nonporphyrin Ni and V.[19,36] The nature of the nonporphyrin metal compounds is not known but several studies of V in crude oils indicate that the metal compounds are associated with the high-molecular-weight asphaltenes.[27,18] Sugihara, *et al.*[18] have noted that nonporphyrin Ni and V are associated with the high-molecular-weight fraction of the Boscan asphaltenes and are absent from the low-molecular-weight fraction.

Metalloporphyrins were sought in the oil C-1, which contains relatively high concentrations of Fe, Zn, Hg, Co, Cr and Ni. Examination of the absorption spectra of the methanol-soluble, resin, and asphaltene fractions of the oil showed the presence of porphyrin complexes in all fractions. Liquid chromatography on silica and alumina was used to separate the metalloporphyrins in the low-molecular fractions of the three oil components. Only Ni porphyrins and very small amounts of V porphyrins were identified in the chromatographic fractions. The absorption spectrum of the separated Ni porphyrins corresponds to Ni deoxophylloerythroetioporphyrin.[29] The V porphyrin was identified as VO deoxophylloerythroetioporphyrin. No metalloporphyrins other than the Ni and V complexes were identified spectrophotometrically, but the possibility exists that small amounts of Zn, Fe, Co, Hg, Cr, and Cu porphyrins escaped detection. The low-molecular-weight fraction of the asphaltenes contained 480 µg/g Fe, 72 µg/g Hg, 112 µg/g Zn, and 2.67 µg/g Co. These concentrations are sufficiently high to make the detection of porphyrin complexes, if present, relatively easy.   The spectrophotometric detection limit for Ni porphyrins was calculated to be 10^{-4} mol/g which corresponds to 6 x 10^{-3} µg Ni/g.   The sensitivities for the other metal complexes should be similar so that if metalloporphyrins other than those containing Ni and V are present in the low-molecular-weight fraction of the asphaltenes they must account for less than 0.1% of the total metal content.

The Ni porphyrin contents were calculated from the total metalloporphyrin contents (Ni + V) by assuming a Ni/V ratio of 12:5.   The concentrations of the nonporphyrin Ni in the various oil fractions were calculated from the Ni porphyrin contents and the total Ni content of the fractions.   The distribution of Ni as Ni porphyrin and nonporphyrin Ni in the components of the crude oil is shown in Table 2.11. The concentrations of Ni in the GPC-separated molecular weight fractions of the different components of oil RF-1 are shown in Table 2.12.   The data in Table 2.11 indicate that 66% of the Ni in the crude oil occurs as Ni porphyrin.   In the methanol-soluble fraction all the Ni occurs as Ni porphyrin.   For the resins and asphaltenes the Ni porphyrin percentages are 64% and 49.2% respectively.   The results presented in Table 2.12 show that the Ni porphyrin contents of the low-molecular-weight fractions of both asphaltenes and resins account for all the Ni in these fractions. In both the resins and the asphaltenes the percentage of Ni present as nonporphyrin Ni increases as the molecular weight of the fraction increases.   In the highest-molecular-weight fraction of the oil (fraction 4 of the asphaltenes) the nonporphyrin Ni content accounts for 72.1% of the total Ni.   Similar results were obtained for V in Boscan asphaltenes by Sugihara, et al.[18]   The data for the elements Co, Fe, Zn, Cr and Hg suggest that metalloporphyrins of these elements are absent and that their behavior is similar to that of the nonporphyrin Ni.

Origin of Metal Compounds

The metals bound in the organic matrix of crude oils may have originated in several ways:

(a)   through incorporation and diagenesis of metal complexes of the original biological material

(b)   through incorporation into the organic matrix during diagenesis of the biological material in the source rocks either from clay minerals or interstitial aqueous solution

Table 2.11

Distribution of Ni and Ni Porphyrin in Crude-Oil Fractions

Fraction	Crude Oil (%)	Ni Concentration (μmol/g)	Ni Porphyrin (μmol/g)	Ni as Ni Porphyrin (%)	Nonporphyrin Ni (%)
Crude oil	100.0	1.590	1.050	66.0	34.0
Methanol soluble	57.5	0.123	0.142	100.0	0
Resins	37.5	2.500	1.600	64.0	36.0
Asphaltenes	4.99	14.500	7.130	49.2	51.8

Table 2.12

Distribution of Ni and Ni Porphyrin in Asphaltenes and Resins of C-1

Molecular Weight Fraction (GPC)	Asphaltenes or Resins (%)	Ni Concentration (μmol/g)	Ni Porphyrin (μmol/g)	Ni as Ni Porphyrin (%)	Nonporphyrin Ni (%)
Asphaltenes					
1	11.2	22.6	23.10	100.0	0
2	23.2	3.2	1.56	49.0	51.0
3	50.6	16.8	5.80	34.5	65.5
4	15.2	18.1	5.04	22.9	72.1
Total	100.0	14.5	7.13	49.2	50.8
Resins					
1	29.4	3.50	3.79	100.0	0
2	21.2	1.87	0.44	38.6	61.4
3	49.4	1.36	0.72	32.0	68.0
4	0	--	--	--	--
Total	100.0	2.50	1.60	64.0	36.0

    (c)   through uptake from an aqueous phase or mineral
         phases during primary or secondary migration
    (d)   from formation waters or from reservoir rock
         minerals.

Most authors have concluded that the metals in petroleum were derived from the primary biological material. Metal-organic complexes derived from biological material have been shown to occur in many types of sediments[37] but it should also be noted that a non-biological origin of the metals in the organic matter of sediments is also likely as indicated by the high sequestering ability of such compounds as humic acid for metal ions.[38] Hodgson[6] in his study of V, Ni and Fe in West Canada oils deduced that the metals entered the oil early in its history, and that the original organic matter was the source of the metals. He considered that loss or gain of metal during migration was unlikely. Hyden,[7] however, concluded that only Ni and V reflect the composition of the organic material, and that other elements (*e.g.*, U, Cu) were introduced from the source rock, the reservoir rock, or during migration. Gulyaeva and Punanova[39] have studied the distribution of 25 elements in Russian petroleum ashes and compared the data with the concentrations of these elements in living organisms and argillaceous rocks. They showed that the petroleum concentrations of Si, Al, Fe, Ca, Mg, Na, Ti, Ba, Mn, Sr, Zn, Cu, Sn, As, Br and Mo are much lower than in marine organisms, that the B, Cr, Pb, Ag, Ni, Co and I concentrations of petroleum and marine organisms were similar, and that V occurred in petroleum at much higher concentrations than in organisms. They concluded that, except for V, living organisms could account for the concentrations observed in petroleum. A similar conclusion was reached by Kasymov.[40] Although all of the metals present in petroleum do occur in living organisms it is difficult to explain the great differences in trace element suites observed in oils without attributing the variation to different conditions of petroleum genesis.[15]

    The observation that the trace elements observed in the California oil occur in oil-soluble form and are predominantly associated with the asphaltic component of the oil (except As) suggests that the origin of the trace elements is related to the origin of the asphaltenes and related resins. The fact that both Ni and V occur in at least two forms, porphyrin and nonporphyrin, suggests that two distinct processes may have occurred. Theories of the origin of porphyrin

complexes in petroleum have been summarized by
Hodgson, *et al.*[19] If porphyrins originated through
degradation of chlorophylls with loss of labile Mg
followed at a later time by metallation with Ni and
V, then the source of Ni and V may have been the
source-rock organic material or the nonporphyrin Ni
and V present in the asphaltenes. The work reported
here has shown that the trace elements Fe, Cr, Zn,
Cr, Hg, Cu are similar in behavior to nonporphyrin
Ni and V. Yen, *et al.*[17] have suggested that nonporphyrin Ni and V occupy "holes" bordered by S, N,
or O atoms in the asphaltene sheets. Erdman and
Harju[19] have shown that asphaltenes are undersaturated
with metal ions and that sites in the sheets are available for metal-ion complexing. They showed that V,
and to a lesser extent Ni and Cu, was taken up from
aqueous solution by asphaltenes. The authors concluded that coordinating sites were formed slowly
during asphaltene genesis, and while the oil was
dispersed in the source rock these sites would be
rapidly filled by Ni^{2+} and VO^{2+} ions. Migration
could also result in metal uptake, in which case the
metals incorporated into the asphaltene structure
would reflect the mineralogical and chemical composition of the migration pathways rather than that
of the source organic material.

Uptake of metals other than V, Ni and Cu was
not considered by Erdman and Harju[19] but complexing
of Fe, Zn, Co, Cr, and Hg would be possible as these
elements all form complexes with N, S, or O-containing
ligands. Gulyaeva and Lositskaya[41] studied uptake of
V (V) by petroleum from aqueous solution. They showed
that uptake would only occur when the oil contained
asphaltene and that it was inhibited by high concentrations of NaCl, indicating that formation waters
could not be the source of V in petroleum. It should
be noted, however, that these experiments were carried out with V (V) and not with V (IV), the form
present in both porphyrin and nonporphyrin V in oil.
Lositskaya and Gulyaeva[42] showed that Ni and Zn were
taken up by petroleum from aqueous solution, but they
inferred that napthenic acids were responsible and
not ligands in the asphaltenes although evidence was
not presented. In a recent study in this laboratory[43]
asphaltenes were shown to absorb up to 300 ppm Cu (II)
from aqueous solution, confirming the results of
Erdman and Harju.[19]

In uptake of metals from aqueous solution,
equilibration with an aqueous phase and adsorption
of metal ions on mineral surfaces may be important.
Brindley[26] has discussed the role of organic molecules

in metal complex formation on clay mineral surfaces, and this mechanism may be important in incorporation of metals in asphaltenes. The mechanism proposed by Erdman and Harju[19] would also explain the increased concentration of Fe, Zn and Hg in the 300-1000 molecular-weight fraction because in the maturation process the degradation of the asphaltenes would lead to simpler organometallic molecules (porphyrins or other complexes). Only Ni and V porphyrins would be likely to survive due to their high thermodynamic stability, but other complexes of Fe, Zn and Hg, which would not show porphyrin spectra, might be formed. These complexes would eventually be decomposed, leaving only Ni and V porphyrins as the metal species in highly paraffinic oils. The possibiltiy of uptake of metals by asphaltenes in the reservoir rock also exists. Indirect evidence for this is provided by the anomalously high concentration of Mo, Dy, and Eu in the Athabasca tar sand oil, which may be related to the presence of minerals, such as molybdenite in the sand.[45]

The incorporation of Sb and As into the asphaltenes may involve a different mechanism from that of the metals as these elements may replace S in the asphaltenes. Degradation of the asphaltenes would possibly reduce alkyl or aryl arsines R_xAsH_{3-x} or stibines, R_xSbH_{3-x}, several of which are water soluble.

CONCLUSIONS

The following conclusions may be drawn from this study:

1. Oil-soluble compounds of Ni, V, Fe, Co, Cr, Hg, Zn, As, Sb, Cu and Na are present in a California Tertiary crude oil.

2. Nickel is present in the oil as Ni porphyrin and nonporphyrin Ni. Nickel porphyrin was found in all fractions of the crude oil and accounted for 100% of the Ni in low-molecular-weight fractions of the three oil components. The proportion of nonporphyrin Ni and the resins and asphaltenes increased with molecular weight.

3. The elements Fe, Co, Zn, Hg, Cr, and Cu occur in the oil as a nonporphyrin form similar to that of Ni and V. Porphyrin complexes of these elements, if present, were not observed.

4. The elements Fe, Co, Hg, Zn, Cr and Cu apparently are incorporated into the asphaltene sheet structure through complexing at "holes" bordered by atoms of N, S, or O.

5.   Arsenic and antimony appear to be present partly as low-molecular-weight compounds such as alkyl or aryl arsines and stibines.   In the case of Sb, these compounds are strongly associated with the asphaltenes.

6.   The origin of the Fe, Co, Hg, Zn, Cr, and Cu in the asphaltenes may involve complexing from an aqueous or solid phase during maturation of petroleum in the source rocks or during migration.

The question of how and when trace metals were introduced into petroleum is far from resolved and the problem presents some extremely interesting avenues for future research.   When more information on such mechanisms is available, the use of trace element data in petroleum geochemistry will rest on a sounder theoretical basis.

REFERENCES

1.   Hackford, J. E.   *J. Inst. Petrol. Technol.*, *8*, 193 (1922).
2.   Ramsay, A.   *J. Inst. Petrol. Technol.*, *10*, 87 (1924).
3.   Shirey, W. B.   *Ind. Eng. Chem.*, *23*, 1151 (1931).
4.   Bonham, L. C.   *Bull. Amer. Assoc. Petrol. Geol.*, *40*, 897 (1956).
5.   Erickson, R. L., A. T. Myers and C. A. Horr.   *Bull. Amer. Assoc. Petrol. Geol.*, *38*, 2200 (1954).
6.   Hodgson, G. W.   *Bull. Amer. Assoc. Petrol. Geol.*, *38*, 2537 (1954).
7.   Hyden, H. J.   U.S. Geol. Survey Bull. 1100-B (1961).
8.   Ball, J. S., W. J. Wenger, H. J. Hyden, C. A. Horr and A. T. Myers.   *J. Chem. Eng. Data*, *5*, 5533 (1960).
9.   Al-Shahristani, H. and M. J. Al-Atiya.   *Geochim. Cosmochim. Acta*, *36*, 929 (1972).
10.   Demenkova, P. Y., L. N. Zakharenkova and A. P. Kurtalskaya.   *Tr. Vses. Neft. Nauk. Issled. Geol. Inst. No. 123*, 59 (1958).
11.   Gilmanshin, A. F., M. G. Gazinov and E. N. Baturina.   *Tr. Tater. Neft. Nauk. Issled. Inst. No. 18*, 113 (1971).
12.   Kotova, A. V., L. N. Tokareva and V. G. Berkovskii.   *Tr. Inst. Khim. Prir. Solei. Akad. Nauk. Kuz. SSR No. 1*, 83 (1970).
13.   Katchenkov, S. M. and E. I. Flegentova.   *Vestsi. Akad. Navuk. Belarus. SSR. Ser. Khim. Navuk.*, 95 (1970).
14.   Botneva, T. A.   *Chem. Abstr.*, *78*, 138768 (1973).
15.   Mileshina, A. G., S. A. Punanova and N. M. Chekhovskikh   *Geol. Neft. Gaza*, *15*, 41 (1971).
16.   Nurev, A. N. and Z. A. Dzabharova.   *Issled. Obl. Neorg. Fiz. Khim.* (1970), p. 33.

17. Yen, T. F., T. G. Erdman and A. J. Saraceno. *Preprints, Div. Petrol. Chem. ACS, 6 (3),* B53 (1961).

18. Sugihara, J. M., J. F. Branthaver, G. Y. Wu and C. Weatherbee. *Preprints, Div. Petrol. Chem. ACS, 15* (2) C5 (1970).

19. Yen, T. F. *Preprints, Div. Petrol Chem. ACS, 17* (4) F120 (1972).

20. Colombo, V. P., G. Sironi, G. B. Fasalo and R. Malvano. *Anal. Chem. 36,* 802 (1964).

21. Patek, P. and H. A. Bildstein. *Anal. Chem., 231,* 187 (1967).

22. Shah, K. R., R. H. Filby and W. A. Haller. *J. Radioanal. Chem., 6,* 185 (1970).

23. Shah, K. R., R. H. Filby and W. A. Haller. *J. Radioanal. Chem., 6,* 413 (1970).

24. Filby, R. H. and K. R. Shah. Proc. American Nuclear Society Topical Meeting Nuclear Methods in Environmental Research, University of Missouri,Columbia, Mo., August 23-24, 1971, p. 86.

25. Hitchon, B., R. H. Filby and K. R. Shah. *Preprints, ACS Div. Petrol. Chem., 18,* 623 (1973).

26. Filby, R. H. and K. R. Shah. *The Role of Trace Metals in Petroleum,* T. F. Yen, Ed. (Ann Arbor, Mich.: Ann Arbor Science Publishers, 1974).

27. Blumer, M. and W. D. Snyder. *Chem. Geol., 2,* 35 (1967).

28. Sugihara, J. M. and R. M. Bean. *J. Chem. Eng. Data, 5,* 106 (1960).

29. Hodgson, G. W., B. L. Baker and E. Peake in *Fundamental Aspects of Petroleum Geochemistry,* B. Nagy and U. Colombo, Eds., (Amsterdam: Elsevier, 1967), Chapter 5.

30. Hodgson, G. W., E. Peake and B. L. Baker. *Research Council of Alberta Information Series 45,* 75 (1963).

31. Dunning, H. N. and N. A. Rabon. *Ind. Eng. Chem., 48,* 951 (1956).

32. Witherspoon, P. G. and R. S. Winniford in *Fundamental Aspects of Petroleum Geochemistry,* B. Nagy and U. Colombo, Eds. (Amsterdam: Elsevier, 1967), Chapter 6.

33. Moore, J. W. and H. N. Dunning. *Ind. Eng. Chem., 47,* 1440 (1955).

34. Falk, J. E. *Porphyrins and Metalloporphyrins.* (Amsterdam: Elsevier, 1964).

35. Hodgson, G. W., J. Flores and B. L. Baker. *Geochim. Cosmochim. Acta, 33,* 532 (1969).

36. Dunning, H. N., H. Baker, R. B. Williams and J. W. Moore. *Preprints, Div. Petrol. Chem. ACS, 5* (1) 169 (1960).

37. Saxby, J. D. *Rev. Pure and Appl. Chem., 19,* 131 (1969).

38. Stevenson, F. J. and M. S. Ardakani in *Micronutrients in Agriculture,* J. J. Mordtvedt, Ed. (Madison, Wis.: American Society of Agronomy), Chapter 5.

39. Gulyaeva, L. A. and S. A. Punanova. *Izv. Nauk. SSSR Ser. Geol. 1973,* 112 (1973).

40. Kasymov, K. K.   *Vop. Geol. Neftgazonos. Uzb. No. 1,* 35
     (1966).
41. Gulyaeva, L. A. and I. F. Lositskaya. *Geochemistry Inter-
     national* (Eng. tr) *1967 (7),* 699 (1967).
42. Lositskaya, I. F. and L. A. Gulyaeva. *Uch. Zap. Azerb.
     Gos. Univ. Ser. Khim. Nauk. No. 4,* 44 (1966).
43. Shah, K. R.   Personal communication.
44. Brindley, G. W.   *Reunion-Gelga de Minerales de la Arcilla,*
     (Madrid, 1970), p. 55.
45. Filby, R. H. and K. R. Shah. Unpublished results.

CHAPTER 3

DETERMINATION OF TRACE METALS IN PETROLEUM
GENERAL CONSIDERATIONS

James Eppolito and H. A. Braier
Gulf Research & Development Company
Pittsburgh, Pennsylvania

GENERAL CONSIDERATIONS ON TRACE ANALYSIS

Introduction

In recent years analytical methods have increased sensitivity, thereby developing a growing
awareness of the significance of trace elements and
the importance of detecting them accurately. The
demands of research and the availability of modern
instrumentation led to the development of new and
specialized methods for the measurement of elemental
composition at the ppm or ppb levels. Thus the distinctive field of trace analysis emerged within the
realm of analytical chemistry.
    The main feature of trace analysis is not the
determination of a minute quantity of a substance,
which in general is not a difficult task, but the
determination of a small quantity in the presence of
an overwhelming amount of other substances. These
other substances, which are generally part of the
sample matrix, may seriously affect the behavior of
the trace constituent. A trace is generally and arbitrarily accepted as a constituent present in samples
below a concentration of 100 parts per million.
    The modern approach to elemental trace analysis
is by means of physical methods. These methods are
generally fast and require small samples that may
need little or no processing. In most cases the
elements are determined regardless of their oxidation
state and chemical combination, and usually many elements can be determined simultaneously. In general,
physical methods produce some permanent record on

photographic film or strip chart, which can be used
for future reference or verification. But there
are no universal methods. Every technique has its
peculiar advantages and limitations.

SELECTION OF AN ANALYTICAL METHOD

The solution of a trace analytical problem lies
in the selection of the analytical approach that will
produce the quality of information required and the
limitations of which are unimportant to the solution
of that particular problem. Considerable compromising
is involved in the selection of an appropriate analytical technique.

Fundamental requirements such as the sensitivity
and accuracy needed must be clearly established before looking for a solution to the problem. Sensitivity refers to the ability to see a small change
in concentration. A closely related term is "detection limit," which is the lowest concentration that
can be determined with a given degree of confidence.
Accuracy refers to the reliability of the result, or
how close it is to the true value. Precision refers
to the repeatability of the result. Obviously it is
possible to have good precision with bad accuracy
but good accuracy implies good precision.

After one or more methods have been selected on
the basis of their sensitivity and accuracy, its
specificity must be established as satisfactory.
Specificity refers to the ability of measuring one
element with no significant effects from other elements present in the system. In general the condition
under which a method is specific are met by proper
regulation of the chemical system and instrumental
parameters. However, due to the nature of certain
samples, these modifications cannot always be successfully applied.

Caution should be exercised when using information available in the technical literature about the
terms described above. Terms such as *sensitivity, precision* and *specificity,* as generally given by instrument
manufacturers, are measured when the element being
determined is present in a pure solvent. This is an
ideal situation, far from what is found in real samples. As a rule of thumb, limits of detection given
by instrument manufacturers should be multiplied by a
factor of 2 to 10. The same applies to analysis time.
In 1933, G. E. F. Lundell[1] published a paper entitled
"The Chemical Analysis of Things as They Are" in

which he complained that many talks and articles on
analytical matters dealt with "the chemical analysis
of things as they are not." The situation has not
changed.

The following are considerations of variable
importance in the selection of an instrumental method.
Is the instrument available in our laboratory? If
not, how much does it cost? What kind of space and
utilities does it require? Can this instrument determine one element at a time or various elements
simultaneously? How is the sample to be presented
to the instrument? Will our particular type of
sample require processing? Can the instrument be
operated by a technician? Can the data be directly
transformed into concentration or must it be processed or interpreted? Useful information concerning
reliability, frequency of adjustments, calibration
and repairs can be obtained from users whose names
are usually provided by the manufacturer. And finally
an often forgotten consideration: is there in town a
factory representative qualified to service the instrument?

PRECAUTIONS IN TRACE ANALYSIS

Contamination

Contamination in trace analysis may be a serious
source of error, especially in the determination of
common and widespread elements. A specimen exposed
to atmospheric dust, for example, may gain as much
sodium as was originally present. Contamination may
occur from the moment of sampling to the last stage
of the analytical procedure. It may be introduced
by unclean containers, laboratory ware, reagents,
atmospheric dust or distilled water. Even dandruff,
perspiration, fingerprints and cosmetics are known
to cause contamination in the trace analysis of some
elements. If the sample must be chemically processed,
a properly run blank will correct for contaminants
contributed by laboratory ware and reagents. However,
contamination experienced during sampling, transfer
and storage will affect the sample before its analysis
and will not be accounted for by the blank.

Sampling

Sampling, which normally precedes analysis, should
be carefully planned and executed because its quality
regulates the value of everything that follows.

Generally a laboratory sample is removed from the bulk product to be analyzed, and this sample is then divided into analytical subsamples. The quantitative result generated from an analytical subsample is limited by how well the subsample represents the bulk. Of course, the bulk may be only a portion of a larger volume.

Trace elements are generally not uniformly distributed in solids or nonhomogeneous materials like most fuel oils. Trace elements of interest in a fuel oil may be associated with the particulate matter or sediment, with the solution, or both. Highly sensitive instrumental techniques often require a very small specimen, thereby increasing the danger that the specimen being analyzed may not constitute a representative portion of the original sample.

Sample and Standard Storage

Standards and laboratory samples are usually stored for variable periods of time. During storage, precautions must be taken to minimize changes such as particulate segregation, precipitation, oxidation, polymerization and other slow reactions. Exposure to light, heat, air and vibration should be avoided. Plastic containers should not be used to store liquid samples, especially if they are of organic nature. Containers should be tightly closed to avoid the evaporation of light components that will leave a concentrated sample. Also, trace metals can be adsorbed on the walls of the containers or contamination may occur from desorption of previously adsorbed contaminants.

Sample Processing

As mentioned previously, many instrumental methods can handle a sample as it is. Sometimes processing is called for to comply with standardized procedures or to create conditions for the specific determination of a substance. In trace analysis, sample processing is performed to concentrate a trace constituent to a level at which it can be determined by a given instrumental technique or to eliminate interfering constituents, or both. Some of the techniques used are solvent extraction, evaporation, distillation, precipitation and ashing.

GENERAL CONSIDERATIONS FOR DETERMINATION   63

EVALUATION OF EXPERIMENTAL DATA

Finally there is the task of evaluating the
experimental data through the use of statistics.
The reliability of an experimental observation must
be measured by comparing it to an estimate of its
error.   This is referred to by statisticians as a
"test of significance."

The first approach for the evaluation of a body
of observations is the calculation of the simple
mean.   About this mean there can be calculated the
average deviation, a figure offering a false sense
of security.   It is not a very accurate measure of
precision, giving a bias measurement and appearing
to be more precise than it actually is.

Another approach is the calculation of the
variance and the standard deviation about the average.   These are the most efficient measurements for
precision, upon which are based all other statistical
analyses.

The principal reason for making an elemental
analysis is to estimate the true value of the amount
of a trace metal or element.   Every refined analytical
method has an inherent error; hence the precision of
the analysis can be improved by replication.   The
efficiency of the mean so determined increases with
the number of replicates.

There exists an interval on each side of the
mean wherein lies this true value.   This interval is
referred to as the confidence limits, and its width
is determined by (a) the number of replicates,
(b) the variability of the analytical method, and
(c) the probability level desired.

If we wish to compare the precision of two different methods for determining a given element, or
a method after it has been modified, we employ the
variance ratio statistic, called the F distribution.
The ratio of the larger variance over the smaller
is compared with a critical value at the desired
probability as a test of significance.

The test of significance for several variances,
whereby an analytical method is applied to several
samples or where several methods are to be compared
after analyzing the same sample, is generally known
as Bartlett's test.   Specifically we wish to know
whether the several variances could reasonably exist
in the same population.

The last statistical technique to be mentioned
is the chi-square distribution.   Is the drift in a
piece of electronic apparatus, for example, greater

than is expected? Observing the ratio of the square of deviations from the mean of a measured element, and the square of the population standard deviation, we can accept or reject the hypothesis that the data and expectations are consistent. This could lead to a study on the influence of apparatus effect on the data quality.

CONCLUSION

   In chemical analysis, and especially in trace analysis, every step, no matter how simple, is important, from sampling to the writing of the final report. Each operation contributes certain variabilities to the final result. A completed analysis can properly be compared to a chain, in the sense that no chain is stronger than its weakest link.

REFERENCES

1. Lundell, G. E. F. *Ind. Eng. Chem., Anal. Ed. 5,* 442 (1933).

CHAPTER 4

DETERMINATION OF TRACE METALS IN PETROLEUM
INSTRUMENTAL METHODS

H. A. Braier and James Eppolito
Gulf Research & Development Company
Pittsburgh, Pennsylvania

INTRODUCTION

As mentioned in the previous chapter, the modern
approach to elemental trace analysis is by means of
physical methods. What follows is a review of the
general principles of some instrumental methods that
can be used to determine trace metals in petroleum.
It is emphasized that there are no universal methods;
each one exhibits its own advantages and disadvantages, which are listed for each technique described.
Some who have worked with these techniques may
disagree with what, in the context of this presentation, is considered to be advantageous or detrimental.
Of necessity, those who have worked for a long time
in a given analytical field will sometimes favor
their familiar approach at the expense of other good
possibilities. The writer of this paper does not
claim to be different and will not be surprised if
he too is found in the predicament just described.

NEUTRON ACTIVATION ANALYSIS

Neutron activation is a well-established method
of elemental analysis capable of attaining very low
limits of detection for most elements. It is not
as widely used as other methods because large instal-
lations or complex machines are needed to achieve
low limits of detection, although this is now changing
with the advent of the californium-252 isotopic neutron sources.

Advantages

There is little or no sample preparation and
there is a very high sensitivity for most elements.
Interferences, if observed, can be eliminated by
instrumental or chemical means or by the application
of proper corrections. Analysis time can be very
short, but this depends on the element to be determined and other factors. Neutron activation analysis
on a service basis is available from a number of
private and government laboratories throughout the
country.

Disadvantages

Because of the use of radiation, special licens-
ing, training of personnel and controls are required.
To achieve high sensitivities, high neutron fluxes
are required, which are expensive.

Cost and Space Needed

Cost and space needs can vary widely. About
$35,000 and a medium-sized room are needed for a
one milligram californium-252 source system. Several
hundred thousand dollars and a building are needed
for a nuclear reactor. Particle accelerators are of
intermediate cost and space requirements.

Principle

In neutron activation analysis, a sample is
irradiated or bombarded with neutrons. Each atom
that captures a neutron undergoes a nuclear reaction
and could be transformed into an unstable or radioactive element. This unstable element, called
radionuclide, becomes stable by releasing energy in
the form of radiation. Each radionuclide is characterized by its rate of radioactive decay or half
life, the type of radiation emitted, and the energy
of the radiation. These characteristics, like nuclear fingerprints, unambiguously identify the arti-
ficially created element, and by inference the
element in the sample that gave origin to it.
Once the sample has been irradiated, two main
approaches can be followed: chemical separation
or instrumental analysis. Chemical separation,
which requires sample processing after irradiation,
is time-consuming and can be quite involved but it

provides the lowest limit of detection afforded by activation analysis. Instrumental activation analysis, which is widely practiced because it is fast and nondestructive, involves the use of gamma-ray spectrometry to study the gamma radiation emitted by the activated sample.

Instrumental neutron activation lends itself to on-stream applications. Figure 4.1 shows a schematic representation of on-stream and discrete sample instrumental activation analysis. In the on-stream example the sample flows through an irradiation coil and then through a counting coil. In the discrete sample application, the sample, in a suitable container, is first irradiated and then taken close to the radiation detector. The radiation detection-analyzing system shown here is similar to the one described for energy dispersive X-ray spectrometry.

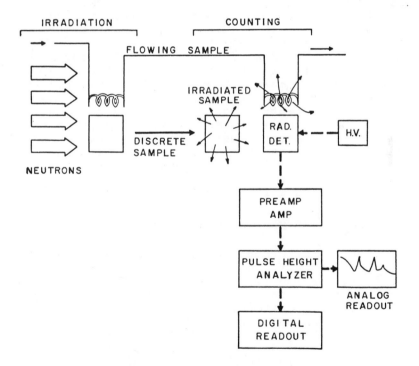

Figure 4.1   Discrete sample and on-stream neutron activation analysis.

Neutron Sources

Neutrons are easy to produce. Being electrically
neutral they easily invade the nuclear domain. For
these reasons and because they react with most elements, neutrons are almost exclusively used in activation analysis. Neutrons can be obtained by means
of nuclear reactors, particle accelerators, and from
isotopic sources.
Nuclear reactors are large and expensive to
build and maintain but they produce high neutron
fluxes that facilitate the attaining of low limits
of detection. Service irradiations and analysis can
be obtained from a number of government-, industry-
and university-owned reactors.
Particle accelerators are machines in which
high speed charged particles collide with suitable
target materials, inducing nuclear reactions with the
production of neutrons or other particles. Van de
Graaff and Cockroft-Walton machines are typical of
this group. The complexity of some of these machines
can be realized by inspecting Figure 4.2, which shows
the 3-Mev Van de Graaff electron accelerator installation at Gulf Research and Development Company.
Notice the four large rooms needed and the five-foot
thick solid concrete shielding wall.
Isotopic neutron sources have the advantages of
compactness, simplicity and lack of moving parts.
Also their neutron output is uniform, stable and
predictable. These characteristics make them ideal
for on-stream applications. Until a few years ago
the known sources of this type produced low neutron
fluxes unsuitable for most applications.

Californium-252 Neutron Sources

For isotopic sources the picture has changed
with the discovery in 1955 of californium-252, an
artificial radionuclide that decays by spontaneous
fission with an intense neutron production. Figure
4.3 shows the californium-252 source holder used at
Gulf Research and Development Company. It is about
four inches long and can contain several milligrams
of the radionuclide. About one milligram of cali-
fornium-252, surrounded by an appropriate moderator,
will produce approximately the same useful neutron
flux as can be produced with the installation shown
in Figure 4.2. Californium-252 is commercially
available from the United States Atomic Energy Com-
mission.

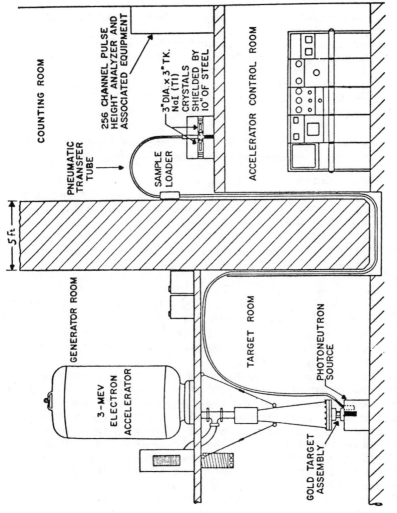

Figure 4.2. Facility for slow neutron activation.

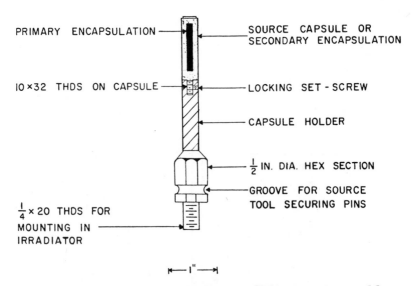

PRIMARY ENCAPSULATION

SOURCE CAPSULE OR
SECONDARY ENCAPSULATION

10×32 THDS ON CAPSULE

LOCKING SET-SCREW

CAPSULE HOLDER

$\frac{1}{2}$ IN. DIA. HEX SECTION

GROOVE FOR SOURCE
TOOL SECURING PINS

$\frac{1}{4}$×20 THDS FOR
MOUNTING IN
IRRADIATOR

|←— 1" —→|

*Figure 4.3.   Californium-252 source holder used at Gulf
Research and Development Company.*

Figure 4.4 depicts an on-stream californium-252
irradiator in which space and radiation shielding
requirements are reduced to a minimum by an under-
ground location. Stream inlet and outlet are at
ground level, while the irradiation coil surrounding
the californium-252 source is about two feet below
ground. A similar analyzer in an above-the-ground
design can be seen in Figure 4.5, which also shows
the counting electronics, readout system, and inlet-
outlet pipes on top of the tank. The three-foot
diameter steel barrel is filled with water for
radiation shielding purposes. These on-stream ir-
radiators can be easily adapted for the analysis of
discrete samples. Figure 4.6 shows a calibration
curve for vanadium in crudes obtained with the above-
the-ground on-stream analyzer just described. The
chart paper readout in Figure 4.7 indicates how
rapidly this analyzer reacts to a sudden change in
vanadium concentration from 2.3 ppm to 1.7 ppm. By
changing operational parameters, vanadium can be
determined from fractional ppm to per cent concentrations.

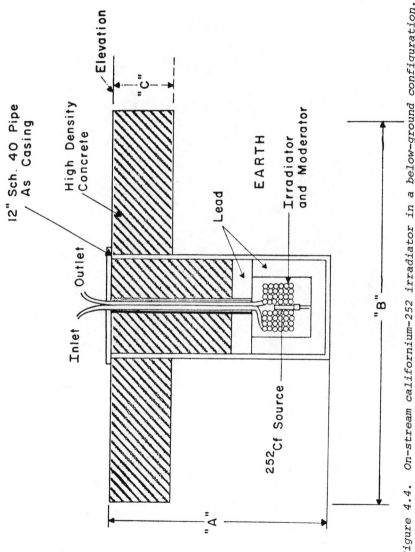

Figure 4.4.   On-stream californium-252 irradiator in a below-ground configuration.

Figure 4.5.   *Above-the-ground on-stream vanadium analyzer featuring a californium-252 source.*

X-RAY FLUORESCENCE

X-ray fluorescence is a well-known and widely used method of elemental analysis. It is directly applicable to solid or liquid samples of organic or inorganic nature.

Advantages

X-ray fluorescence analysis can be performed using rapid and simple techniques with little sample

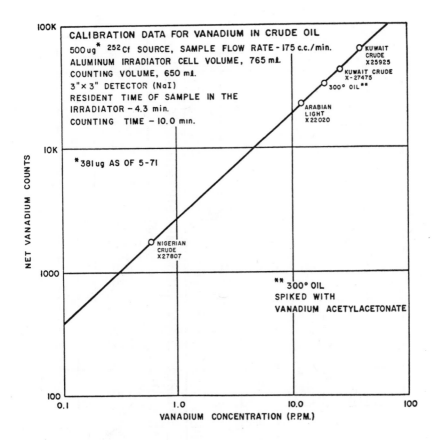

Figure 4.6

preparation for concentrations above 0.01%. It is
an excellent method when applied to routine run-of-
the-mill samples.

Disadvantages

To determine low or fractional parts per million
levels, most samples must be processed to concentrate
the trace elements. The technique is not sensitive
for sodium and lighter elements. Matrix effects are
important and affect accuracy unless appropriate
standards or corrections or both are used.

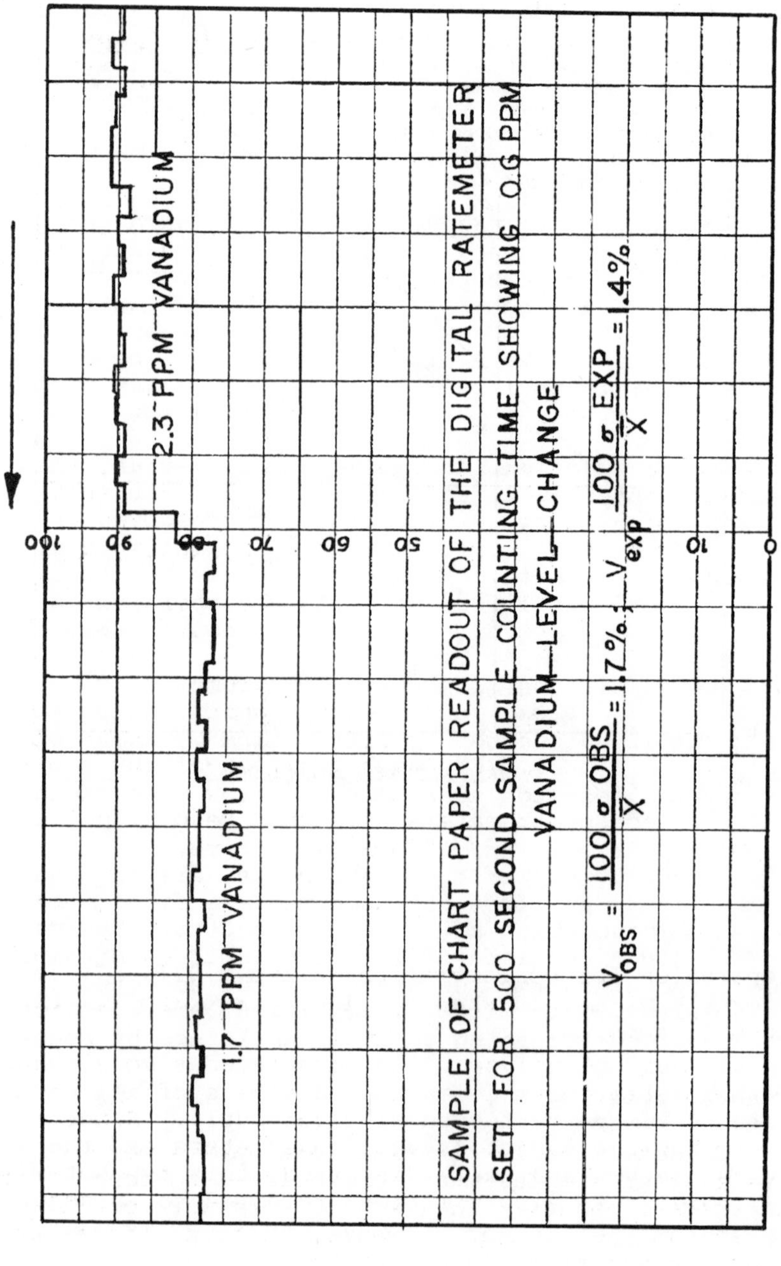

Figure 4.7.   Recorder readout of a vanadium level change.

Cost and Space Needed

The price ranges from $20,000 to $40,000 for
a good instrument. The cost can be easily doubled
by the addition of automatic features, computer operation, etc. For safety considerations it is desirable
to have these instruments in a separate room.

Principle

When atoms are excited by X-ray bombardment,
inner shell electrons are ejected. The vacancies
thus created are immediately filled by outer shell
electrons. These outer shell, energetic electrons
move toward the lower energy inner shells by discrete
jumps from energy level to energy level. The resulting decrease in their energy appears as X-ray photons
with wavelengths that are characteristic for each
element.

When a specimen is bombarded with X-rays, the
elements present will emit their characteristic X-rays.
For analytical purposes these must be resolved and
their intensity measured. This is accomplished with
an X-ray spectrometer of which there are two fundamental types, namely the wavelength dispersive and
the energy dispersive. Figure 4.8 shows a schematic
diagram of these spectrometers.

The wavelength dispersive spectrometer is the
conventional optical instrument in which the photons
are spatially dispersed by diffraction by means of a
crystal goniometer. For each position of the goniometer, the detector sees only a narrow wavelength
band. These spectrometers have good resolution, but
can determine only one element at a time.

In energy dispersive instruments, all the characteristic X-rays from the specimen are first detected
by a high resolution solid state detector. The detector output consists of electrical pulses of amplitude proportional to the energy of the detected photons.
These pulses, once linearly amplified, are electronically sorted according to their amplitude by means of
a pulse height analyzer. An advantageous consequence
of this detection mode is that analysis of all chemical
elements can be performed simultaneously. Since it
eliminates the use of crystal goniometers and the
X-rays are not spatially dispersed, this type of
spectrometer is also sometimes referred to as nondis-
persive.

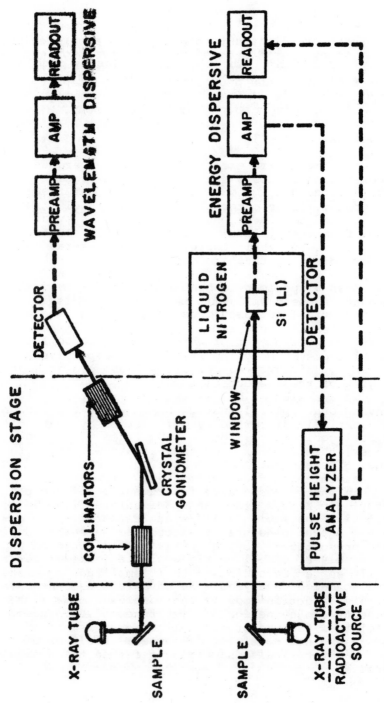

Figure 4.8.   Comparison of wavelength dispersive and energy dispersive X-ray spectrometers.

ATOMIC EMISSION SPECTROSCOPY

There are different techniques in atomic emission spectroscopy that are based upon the types of excitation and detection used. Under this heading arc and spark excitation and photographic and multiphotometric detection will be discussed. Flame photometry, although by principle belonging to this group, will be discussed together with atomic absorption spectrometry.

Advantages

There is qualitative or quantitative determination of about 70 elements, all or most of which can be simultaneously determined. Analysis time can be very short, as is the case with direct readers.

Disadvantages

There is a strong matrix effect and poor sensitivity for volatile elements. Involved calibration is required for precise quantitative analysis.

Cost and Space Requirements

The cost varies from about $30,000 for a relatively simple spectrograph and accessories to $100,000 for more sophisticated instruments. A medium-sized room is the minimum space required to house this equipment.

Principle

In atomic emission spectrosopy the atoms are excited by thermal or electrical means or both. The excited atoms and atomic ions emit their excess energy in the form of light of definite wavelengths. The emitted radiation is resolved by optical means in its individual wavelengths, which are recorded photographically or photoelectrically as a definite pattern of spectral lines. The position of the spectral lines gives a qualitative indication of which elements are present. The intensity of the spectral lines gives a quantitative indication of the amount of each element in the sample.

Figure 4.9 shows the fundamental parts of an emission spectrograph: a means to excite the sample, a dispersive element and a detection system.

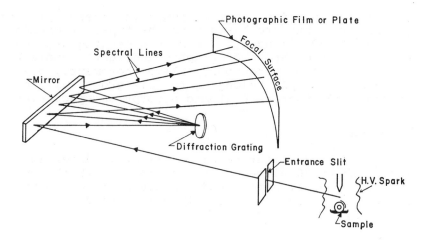

Figure 4.9.   Schematic diagram of an optical emission
              spectrograph.

Sample Excitation

There are three sources commonly used to vaporize
and excite the elements in a sample: the flame, the
electric arc and electric spark discharge.   Flames
possess the lowest excitation energy, producing a
rather simple line spectra.   Flame excitation will be
discussed when dealing with flame photometry and
atomic absorption.

In arc excitation vaporization occurs from the
heating caused by the passage of electrical current.
It produces a complex line spectrum due mostly to
excited neutral atoms.   The DC arc is widely used because it can be applied to almost any kind of solid
sample and because it exhibits the highest sensitivity
for trace element detection.   Its lack of stability
results in somewhat poor precision.

The high voltage AC spark gives higher excitation
energies than the DC arc with much less heating effect.
It produces a complex line spectrum due mostly to excited atomic ions.   Because it is a stable source,
it is preferred when high precision rather than high
sensitivity is required.   It is easily applicable to
liquid samples and solutions.

Dispersion

The most important component of a spectrograph is the dispersive element, which can be either a prism or a diffraction grating. It sorts out by wavelength the radiant energy coming from the excited atoms.

Detection

The dispersed radiation is photographically recorded on film or plate providing a permanent record for analysis. The slowness of the photographic process and subsequent evaluation led to the development of instruments known as direct-reading spectrometers.

Direct-Reading Spectrometers

In a typical instrument, schematically shown in Figure 4.10, the photographic film is replaced with an opaque barrier that has several slits located at wavelengths appropriate to the elements to be analyzed. Behind each slit is mounted a photomultiplier tube, which directly converts radiant energy into electrical energy, permitting a rapid readout. Direct readers are very useful in high speed routine quantitative work. From few to many elements can be determined simultaneously.

ATOMIC EMISSION FLAME PHOTOMETRY AND ATOMIC ABSORPTION FLAME PHOTOMETRY

Although these spectroscopic techniques are conceptually different, they have some points in common and will be discussed together to emphasize their similarities and differences. Both are well-known, widely used, and relatively inexpensive instrumental methods for the determination of metallic elements in organic or inorganic samples.

Advantages

The flame emission technique has a very high sensitivity for some elements and is rapid, simple and inexpensive.
The atomic absorption technique has good sensitivity for most metallic elements, is rapid, simple, and specific, with few interferences.

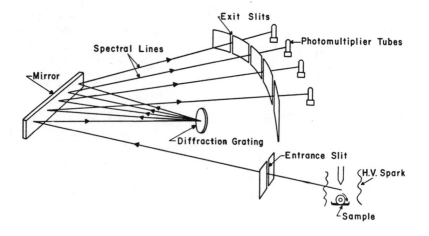

*Figure 4.10.   Schematic diagram of a direct reading
optical emission spectrometer.*

Disadvantages

Flame emission has spectral interferences, and
flame fluctuations affect precision and accuracy.
One element is determined at a time, and the sample
must be in solution.

Atomic absorption can determine one element at
a time. The sample must be in solution, although new
flameless techniques like the carbon rod may change
this requirement.

Cost and Space Needs

Prices range from $5,000 to $15,000 for an instrument that can perform both techniques, and considerably less for a flame emission spectrometer.
Only a few feet of bench space are needed for either
instrument.

Principle

As can be seen in Figure 4.11, both methods use
a burner, a flame, a monochromator, and a detector
that generally is a phototube. A hollow cathode lamp
and a chopper are added for atomic absorption.

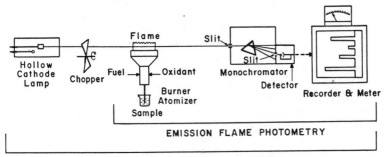

Figure 4.11. Comparison of emission and atomic absorption
flame photometry.

In flame emission as in atomic absorption a
burner aspirates a solution of the sample and atom-
izes it into the flame. The heat of the flame and
the kinetic energy of the combustion products break
down the sample molecules into its component atoms.
Close to 100% of these atoms are not excited but
merely dissociated from their chemical bonds and
placed into a nonionized ground state. An important
property of the ground state atoms of a given element
is that they will absorb light of a specific wavelength.
This light, known as resonant radiation, is the same
that would be emitted by these atoms if they were excited. The resonant radiation of a metallic element
is easily produced with a hollow cathode lamp whose
cathode contains that metallic element. The excitation
energy of the commonly used flames is low compared
to that of arcs and sparks. Consequently less than
1% of the metal atoms in the flame are mildly excited
to emit a simple line spectrum. In both methods, the
monochromator is used to isolate the analytical wavelength used for the analysis from all other unwanted
light produced by the flame or hollow cathode lamp
or both.

The conceptual difference between both methods
is that in flame emission what is measured is the
emission from the excited atoms, which are only a
small part of the metal atoms of a given element present in the flame. In atomic absorption what is measured is the *absorption* of a constant output of certain
resonant radiation by the majority of the metal atoms
of a given element present in the flame, which are
those in the ground state.

In flame emission, spectral interferences are common and affect accuracy. Spectral interferences occur because the different elements excited emit their radiation simultaneously and the monochromator may not have the resolution to isolate an analytical line from the effect of a very intense or a very close interfering line. Precision also suffers because the population of excited atoms is affected by flame fluctuations that are difficult to control. Flame emission is extremely sensitive for the determination of the alkali metals, the alkaline earth metals, and a few other metallic elements.

Due to a combination of factors, spectral interferences are much less prevalent in atomic absorption than in flame emission. For example, by means of a chopper or other suitable device, the continuous output of the hollow cathode lamp is transformed into a modulated analytical signal. This signal is then separated electronically and measured independently from the continuous and interfering signal produced by the flame. Also, the lines from a hollow cathode lamp are few and narrow and are specifically absorbed. Therefore, a given element can only be determined when the lamp and the monochromator setting are both matched to the element analysis. Precision is not significantly affected by flame fluctuations because these have little effect on the population of ground state atoms. Atomic absorption is by no means free of interferences, but they are uncommon and in most cases can be recognized and overcome by the use of standard techniques.

New flameless methods for sample presentation to the instrument, like the graphite rod and the tantalum boat, are expanding the use and applications of atomic absorption spectrometry. In the first technique the burner is replaced by a graphite rod with a small well where a few microliters of sample are deposited and electrically heated by means of a controlled power supply. Important advantages of this procedure are that a very small sample is needed and the dilution of viscous samples is not required. Also, the absence of interfering flame background helps to obtain better sensitivity for many elements. In the boat technique, the liquid sample is deposited in a narrow, boat-shaped container and carefully dried. The container is made of tantalum or other high-melting-point metal. The boat is placed under the light path of the hollow cathode lamp and the dried sample is vaporized by electrical heating or with a conventional acetylene-oxygen flame. This technique, although simple, is not free from interferences and offers better sensitivity for a few elements only.

SPARK SOURCE MASS SPECTROMETRY

This is not a common or well-known technique of elemental analysis. Nevertheless it has reached the point of refined development; instrumentation is commercially available and the technique has attractive features.

Advantages

It is the only method that offers high and uniform sensitivity for all elements with the exception of hydrogen and helium. Typical precisions are ±2-5% above the 1 ppm level and ±10-20% from 1 to 0.001 ppm. All elements can be determined simultaneously.

Disadvantages

It is a sophisticated instrument requiring the services of a professional, and the technique is a relatively slow one. Barring any difficulty, about ten oil samples a day can be analyzed. The instrument is expensive, costing about $120,000 and requiring considerable floor space.

Principle

As the name implies, this is a mass spectrometric technique. The principle of mass spectrometry can be explained briefly with the aid of Figure 4.12, showing the diagram of a double-focusing instrument that is the type used in spark source mass spectrometry.

Mass Spectrometry in General

The mass spectrometer is an instrument that sorts out molecular or atomic ions according to their masses and electrical charges. Several methods of producing mass spectra have been devised, and all of them have the following five parts in common: (1) The inlet sample system. This provides the means for sample introduction. It is generally suited for gaseous and liquid samples and is usually heated to help vaporize the sample in the high vacuum of the instrument. (2) The ion source. In this part the gaseous molecules are ionized and fragmented by electron bombardment. (3) The accelerating system, which consists of a series of slits charged at decreasing potentials

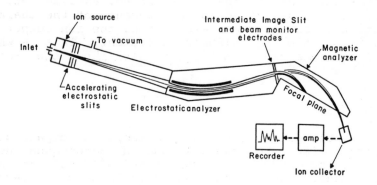

Figure 4.12.   Double-focusing mass spectrometer.

through which the ions formed at the ion source are accelerated. (4) The mass analyzer separates and focuses the ions on a curved or plane surface according to their mass-to-charge ratio. The main difference among the different types of mass spectrometers lies in the design of the mass analyzer. (5) The detection system detects the individual ion beams by means of an ion collector-amplifier-readout system. A photographic plate can also be used to detect and record the mass spectrum. The position of the spectral lines gives a quantitative indication of the amount of each element or molecule present in the sample.

Spark Source Mass Spectrometry

This is a technique adapted to the determination of elemental rather than molecular composition. The main difference between the instrument described above and the kind needed for this technique lies in the design of the sample inlet and ion source. The regular inlet is replaced with a rather large door, which provides access to the ion source area, and the ion source is replaced with electrodes made with the sample to be analyzed. If the sample is liquid, it is frozen on the outside surface of hollow metal electrodes through which liquid nitrogen is constantly circulated. In this manner, a film of frozen sample remains on the electrode's surface even during excitation. A high voltage radio frequency spark, which provides energetic and cool excitation, is used to vaporize and ionize the elemental components of the sample. These

elemental ions are then accelerated, analyzed and
detected as described above. Because of the characteristics of the spark excitation, the ion output
varies little from element to element. This is why
all elements can be determined with about the same
sensitivity.

CONCLUSION

Table 4.1 represents an attempt to compare some
characteristics of the instrumental techniques just
described. It would be unfair and difficult to as-
sign other than general values to these characteristics because they depend on many variables, such
as the nature of the sample, the element to be determined, the particular instrument and perhaps, most
important, the ability of the analyst.

Some of the described instrumental techniques
are complex and uncommon but offer very low limits
of detection. Others are simple and widely used.
Nevertheless, they may become sophisticated and restricted if pushed to give the ultimate detection
limits of which they are theoretically capable.

As new specifications establish lower tolerances,
exotic instrumentation and involved handling, not
available elsewhere, may be needed to analyze petroleum for trace metals. In this regard the investigator should be prepared to send samples for analysis
to laboratories far from his own location. If so,
analytical results should not be expected sooner than
a few days, or in some cases a few weeks.

Table 4.1
Comparison of Instrumental Methods

Characteristic	Neutron Activation Analysis	X-Ray Fluorescence	Emission Spectroscopy	Flame Emission	Atomic Absorption	Spark Source Mass Spectrometry
Scope	most elements	elements heavier than Si	most metals	many metals	most metals	most elements
Sensitivity	ppb, element & neutron flux dependent	ppm, element and matrix dependent	ppm, metal and matrix dependent	ppb for few metals, ppm for many	ppb for few metals, ppm for most	ppb for most elements
Accuracy	good	good	good-poor	good-poor	good	good-poor
Specificity	good	good	good	good-poor	good	good
Contamination	good	good	good	poor	poor	good
Analysis time	fast-slow	fast	fast	fast	fast	slow
Several elements simultaneously	yes	yes: E. Disp. no: W. Disp.	yes	no	no	yes
Instrument cost	$30 K UP	$30 - 50K	$30 K UP	$2 - 4K	$4 - 15 K	$120 K
Space Needed	room-bldg.	room	room	bench	bench	room
Operation	simple to complex	simple	simple	simple	simple	complex
Interpretation	simple to complex	simple to complex	simple to complex	simple	simple	complex
On-stream capability	yes	no	no	no	no	no

REFERENCES

1. Addink, N. W. in *Trace Characterization*, W. W. Meinke and
   B. F. Scribner, Eds., NBS Monograph 100 (Washington, D.C.:
   U.S. Government Printing Office, 1967), pp. 121-147.
2. Ahearn, A. J. in *Trace Characterization*, W. W. Meinke and
   B. F. Scribner, Eds., NBS Monograph 100 (Washington, D.C.:
   U.S. Government Printing Office, 1967), pp. 347-374.
3. Birks, L. S. *X-ray Spectrochemical Analysis* (New York:
   Interscience, 1969).
4. Cornu, A. in *Advances in Mass Spectrometry*, Vol. 4 (New
   York: The Elsevier Publishing Co., 1968), pp. 401-418.
5. Guinn, V. P. in *Treatise on Analytical Chemistry*, I. M.
   Kolthoff and P. J. Elving, Eds. (New York: Wiley, 1971),
   Part I, Vol. 9, Chapter 98, pp. 5585-5641.
6. Kahn, H. L. in *Trace Inorganics in Water*, R. F. Gould, Ed.,
   Advances in Chemistry Series No. 73 (Washington, D.C.:
   American Chemical Society, 1968), Chapter 11, pp. 183-229.
7. Owens, E. B. *Appl. Spectros. 21*(1), 1-8 (1967).
8. Ricci, E. and T. H. Handley. *Anal. Chem. 42*, 378 (1970).
9. Smales, A. A. in *Trace Characterization*, W. W. Meinke and
   B. F. Scribner, Eds., NBS Monograph 100 (Washington, D.C.:
   U.S. Government Printing Office, 1967), pp. 307-326.
10. Williar, H. H., L. L. Merritt, Jr., and J. A. Dean.
    Instrumental Methods of Analysis, 4th Ed. (New York:
    D. Van Nostrand, 1965).
11. Yoakum, A. M. in *Developments in Applied Spectroscopy*,
    Vol. 8, E. L. Grove, Ed. (New York: Plenum Press, 1970),
    pp. 3-17.

CHAPTER 5

NEUTRON ACTIVATION METHODS FOR TRACE ELEMENTS IN CRUDE OILS

R. H. Filby and K. R. Shah
Department of Chemistry and Nuclear Radiation Center
Washington State University
Pullman, Washington

Trace elements in petroleum have received increased attention in recent years because of the value of trace element data in understanding petroleum geochemistry and origin. Data on metal compounds in crude oils are also important to the refinery operator because several elements such as S, V, Cu, and As cause problems in petroleum cracking and refining. In recent years the composition of such fuels as residual fuel oil has been of concern to environmentalists evaluating emissions from industrial concerns (*e.g.*, the presence of vanadium in urban atmospheres from residual fuel burning).[1]

A variety of techniques are available for the determination of trace elements in crude oils, and chemical methods of analysis have been summarized by Milner[2] and McCoy.[3] Most geochemical studies of metals in petroleum have used emission spectrography of petroleum ashes because of the multielement nature of the method.[4-8] In recent years techniques such as polarography,[9-10] colorimetric analysis,[11] X-ray fluorescence,[12] ESR,[13] flame atomic absorption,[14-15] and flameless atomic absorption[16-17] have been used for the analysis of crude oils. Many of the techniques require preconcentration of the metals, usually by ashing techniques and consequently involve the risk of loss of volatile compounds or contamination by reagents. For many elements at very low concentrations (<1 µg/g) the risk of contamination is very high. Most of the applications cited above involve the determination of one or a few specific elements and are not suitable for multielement analysis.

Neutron activation analysis for the determination of trace elements has been used extensively in inorganic geochemistry[18-20] and the technique is well suited to petroleum analysis for the following reasons:

1. The high sensitivity of the method for many elements obviates the use of preconcentration procedures, for example, wet or dry ashing.
2. Small samples may be analyzed because of the high sensitivity of the method.
3. The loss of volatile elements is avoided because samples are not heated above 40°C.
4. Nondestructive multielement analysis is possible.
5. Preirradiation processing is not necessary and extraneous contamination can be reduced to a minimum.
6. Interferences and matrix effects are of minimal importance.

Guinn, et al.,[21] Jester and Klaus,[22] and Braier and Mott[23] have surveyed the potential utility of neutron activation analysis in the petroleum field. A number of investigators[24,25] have used NaI(Tl) spectrometry after radiochemical separation to determine several elements in petroleum. Ordogh, et al.[26] determined I, Br and V in Hungarian crudes by a non-destructive technique, and Palmai[27] used a similar method to determine V, Na, Mn, and Al in products from refinig of crudes. Several Russian papers[28-31] report the use of nondestructive activation analysis and NaI(Tl) gamma-ray spectrometry for such trace elements as V, Ni, Cu, Co and Na. Mast, et al.[32] and Arroyo and Brune[33] have used nondestructive techniques for the determination of V in crude oils. Most studies cited above have not utilized the great potential of neutron activation analysis for multi-element analysis. The introduction of high-resolution Ge(Li) gamma-ray detectors makes instrumental neutron activation analysis (INAA) possible for many elements. Bryan, et al.[34] have used INAA to determine 17 elements in crude and residual fuel oils in a study on fingerprinting crude oils and oil spills. Shah, et al.[35,36] and Filby and Shah[37] have described detailed schemes of INAA for 23 elements, and Filby[38] and Hitchon, et al.[39] have applied these techniques to the trace element geochemistry of petroleum.

The objective of this paper is to describe a comprehensive sequential method for 25 elements in crude oils using high resolution Ge(Li) gamma-ray spectrometry and computer calculation of elemental data. An extension of the method to Cd using radio-chemical separation techniques is described.

EXPERIMENTAL

The analytical method was divided into four steps determined by the half-lives of the radionuclides used for elemental analysis. In step 1 nuclides with short half-lives (less than three hours) are measured to determine V, S, Ca, I, Mn, and Cl. In step 2 nuclides with intermediate half-lives (2.5-36 hrs) and long half-lives (>2 days) are measured for Na, K, Cu, Ga, As, Sb and Br, and for Au, U, Se, Hg, Cr, Zn, Sc, Sb, Rb, Cs, Ni, Co, Fe, Mo, and Sb, respectively. In step 4 Cd is determined by a radiochemical procedure. The nuclear reactions used are shown in Tables 5.1-5.3.

Table 5.1

Nuclear Reactions and Nuclear Properties--Group I

Element	Reaction	$\sigma_{th}.f^a$ (barns)	Product Nuclide Half-life	γ-Ray (keV)
V	$^{51}V(n,\gamma)^{52}V$	4.90	3.75 m	1434.3
Ca	$^{48}Ca(n,\gamma)^{49}Ca$	0.002	8.80 m	3084.4
Cl	$^{37}Cl(n,\gamma)^{38}Cl$	0.106	37.3 m	1642.7
I	$^{127}I(n,\gamma)^{128}I$	6.40	25.0 m	443.3
S	$^{36}S(n,\gamma)^{37}S$	2×10^{-5}	5.1 m	3105.3

a $\sigma_{th}.f$ - product of thermal-neutron capture cross section, σ_{th}, in barns and fractional isotopic abundance, f.

Table 5.2

Nuclear Reactions and Nuclear Properties--Group II

Element	Reaction	$\sigma_{th}.f$ [a] (barns)	Product Nuclide Half-life	Product Nuclide γ-Ray (keV)
Mn	$^{55}Mn(n,\gamma)^{56}Mn$	13.30	2.58 h	846.7
Na	$^{23}Na(n,\gamma)^{24}Na$	0.53	15.00 h	1368.0
K	$^{41}K(n,\gamma)^{42}K$	0.08	12.4 h	1524.7
Cs	$^{133}Cs(n,\gamma)^{134}Cs$	28.0	2.05 y	797.0
Rb	$^{85}Rb(n,\gamma)^{86}Rb$	0.65	18.7 d	1078.8
Cu	$^{63}Cu(n,\gamma)^{64}Cu$	3.10	12.80 h	511.0
As	$^{75}As(n,\gamma)^{76}As$	4.50	26.40 h	559.1
Au	$^{197}Au(n,\gamma)^{198}Au$	97.80	74.70 h	411.8
Mo	$^{98}Mo(n,\gamma)^{99}Mo$	12.10	66.70 h	140.0[b]
Cr	$^{50}Cr(n,\gamma)^{51}Cr$	0.73	27.80 d	320.1
Fe	$^{58}Fe(n,\gamma)^{59}Fe$	0.003	45.60 d	1292.0
Hg	$^{202}Hg(n,\gamma)^{203}Hg$	1.19	46.90 d	279.1
Sb	$^{123}Sb(n,\gamma)^{124}Sb$	1.41	60.30 d	1691.0
Ni	$^{58}Ni(n,p)^{58}Co$	0.10[b]	71.30 d	811.1
Sc	$^{45}Sc(n,\gamma)^{46}Sc$	23.00	83.90 d	889.3
Zn	$^{64}Zn(n,\gamma)^{65}Zn$	0.23	243.00 d	1116.0
Co	$^{59}Co(n,\gamma)^{60}Co$	37.00	5.26 y	1333.0
U	$^{238}U(n,\gamma)^{239}U$	0.05	23.5 m	106.4[b]

a $\sigma_{th}.f$ - product of thermal-neutron capture cross section, σ_{th}, in barns and fractional isotopic abundance, f.

b Daughter-nuclide gamma ray.

Table 5.3

Nuclear Reactions and Nuclear Properties--Group III

Element	Reaction	$\sigma_{th}\cdot f^a$ (barns)	Product Nuclide Half-life	γ-Ray (keV)
Cd	^{114}Cd$(n,\gamma)^{115}$Cd	0.087	53.5 h	528

a $\sigma_{th}.f$ - product of thermal-neutron capture cross section,
σ_{th}, in barns and fractional isotopic abundance, f.

All irradiations were performed in the Washington State University TRIGA Mark-III fueled research reactor at a thermal neutron flux of 8×10^{12} neutrons cm^{-2} sec^{-1}. High efficiency (>15% relative to a 3" x 3" NaI(Tl) detector and 25 cm at 1333 keV) Ge(Li) detectors (FWHM 2.3 keV at 1333 keV at 1333 keV) and a Nuclear Data Model ND 2200 4096-channel analyzer with IBM compatible magnetic tape output were used for gamma-ray spectra measurement. A schematic diagram of the gamma-ray spectrometer is shown in Figure 5.1. Data analysis was performed on an IBM 360/67 computer.

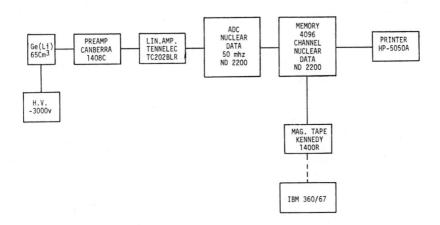

Figure 5.1.   Schematic diagram of gamma-ray spectrometer system.

Procedure

All oil samples were centrifuged and filtered
through Whatman No. 1 or 5μ Millipore filters prior
to analysis. For viscous oils it was necessary to
dilute the oils with benzene before clean-up. The
benzene was removed from the filtered oils by vacuum
evaporation. Polyethylene or quartz vials used for
irradiation were washed with double distilled nitric
acid followed by double-distilled water and acetone
to remove surface impurities.

Step 1

The oil (0.5-1.0 g) was weighed out into a 2/5-
dram polyethylene vial and the vial was heat-sealed.
Aqueous solution standards for V, I, Cl, S, Mn and
Ca were sealed in similar vials. The samples and
standards were irradiated for 15 mins in the reactor
and allowed to decay for 5 mins after irradiation.
Gammy-ray spectra of samples and standards were
measured 5-15 mins after irradiation for ^{52}V, ^{37}S,
and ^{49}Ca and again after 30 mins for ^{128}I, ^{38}Cl and
^{56}Mn. The γ-ray spectrum of a California Tertiary
oil is shown in Figure 5.2.

Step 2

A second aliquot of the oil (1-2 g) was weighed
into a 5-ml quartz vial, which was then sealed in a
2-dram polyethylene vial. Multielement aqueous
solution standards were sealed in 2/5-dram polyethylene
vials.
The standard solutions used contained Na, Br, K;
Se, Ni, Sc, Rb; Cr, Fe, Co, Zn, Cs; and Au, Mo, U,
Sb, Hg, as individual standards. Samples and standards were irradiated for 8 hours in the reactor
and allowed to decay for 24-48 hours. Samples and
standards were prepared for counting by transferring
the sample to a 3"-dia glass petri dish, mixing with
benzene and silica to make a 3" x 0.5" disc. Gamma-
ray spectra were then measured for ^{24}Na, ^{64}Cu, ^{42}K,
^{72}Ga, ^{76}As and ^{82}Br. After 4-10 days γ-ray spectra were measured for ^{198}Au, ^{124}Sb, ^{134}Cs, ^{86}Rb,
^{51}Cr, ^{75}Se, ^{203}Hg, ^{58}Co (for Ni), ^{60}Co, ^{65}Zn, ^{46}Sc,
^{99}Mo, and ^{239}Np (for U). Gamma-ray spectra for the
intermediate half-life group and the long half-life
group are shown in Figures 5.3 and 5.4, respectively.

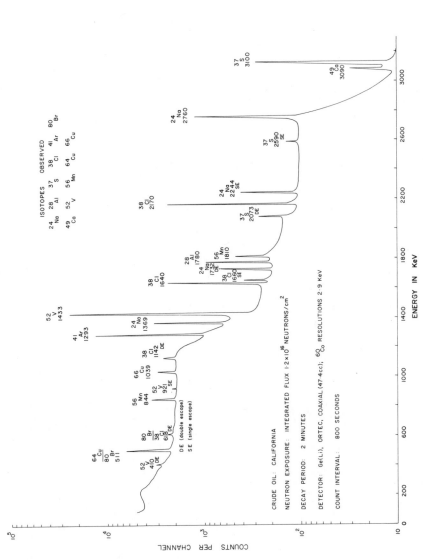

Figure 5.2 Gamma-ray spectrum of neutron-irradiated California crude oil showing short half-life nuclides (step 1).

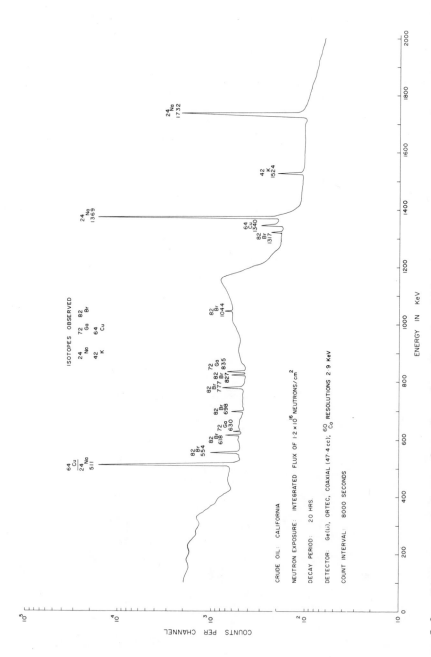

Figure 5.3.   Gamma-ray spectrum of neutron-irradiated California crude oil showing intermediate half-life nuclides (step 2).

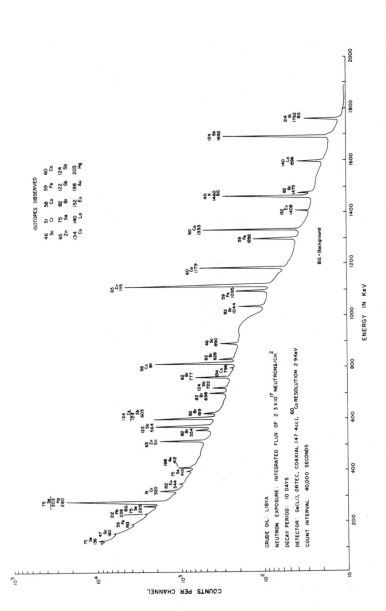

Figure 5.4. Gamma-ray spectrum of neutron-irradiated Libyan crude oil showing long half-life nuclides (step 2).

All γ-ray spectra were collected on 7-track magnetic tape, and the spectra analyzed by FOURIER, a computer program for computing nuclide activities, elemental concentrations and statistical parameters of the analytical results.[40] Corrections for the ^{75}Se interference in the [203]Hg 280 KeV γ-ray were made as described by Filby, *et al.*[41] The ^{64}Cu activity was measured using the positron 511 keV annihilation radiation after correction for the contribution from pair production of the [24]Na 2754 keV γ-ray.

Step 3

An aliquot of the oil (1-10 g) was weighed into a clean quartz vial and sealed in a polyethylene irradiation vial. A solution of $Cd(NO_3)_2$ in 0.1 M HNO_3 containing 35.0 μg Cd/ml was used as a standard. Samples and standards were irradiated for 4-16 hours and allowed to decay for 4-8 hours after irradiation. Each oil sample was transferred to a 200-ml beaker, and the oil remaining in the irradiation vial was then washed out with benzene. The benzene was evaporated off on a hot plate, 25 mg of Cd^{2+} carrier in 15 ml of 18 M H_2SO_4 added and the solution charred for 2 hours on the hot plate until the appearance of SO_3 fumes.

After charring, a 5-ml aliquot of 16 M HNO_3 was added to oxidize organic matter and the solution evaporated to SO_3 fumes. The oxidation step was repeated until a clear solution was obtained. In some cases H_2O_2 oxidation was used after HNO_3 oxidation. When the solution was oxidized, 1 ml of conc. HBr and 5 ml 30% H_2O_2 were added. The solution was then evaporated to dryness. This step removes Sb, As, and Se as volatile bromides.

The residue was dissolved in 5 ml 6 M HCl and the solution was passed through a Dowex-1 anion exchange column. The column was then washed with 25 ml 1 M HCl to remove Fe, Ni, Co, Cu and Na. Zinc was eluted from the column with 10 ml 0.01 M HCl. The Cd was then eluted from the column with 20 ml 0.5 M NH_3 into a 150-ml beaker. The solution was heated almost to boiling, 5 ml of conc. NH_3 and 1 ml of 0.1 M $FeCl_3$ added and the precipitate of $Fe(OH)_3$ allowed to coagulate. The solution was made acid with 6 M HCl and evaporated to approximately 15 ml volume.

The solution was transferred to a 60-ml counting bottle and allowed to stand for 24 hours to establish the ^{115}Cd-^{115m}In equilibrium. When equilibrium was attained the sample was counted for 10,000-40,000 seconds on the γ-ray spectrometer. The irradiated standard solution was quantitatively transferred to the 60-ml counting bottle and diluted with water to give the same volume as that of the sample solution. The 336 keV γ-ray of ^{115m}In in radioactive equilibrium with ^{115}Cd was used to estimate the ^{115m}Cd activity.

RESULTS AND DISCUSSION

Table 5.4 shows values for trace elements in four crude oils selected to show a wide range of trace elements. Similar ranges of values are seen in the data Hitchon, *et al.*[39] obtained using the technique described here. The Cd concentrations of nine crude oils and two asphaltenes are shown in Table 5.5 and the standard deviations listed are based on counting statistics. The detection limits for the technique vary from element to element and are also dependent on the composition of the matrix analyzed (except for Cd). An estimate of practical detection limits has been made by taking a "typical" crude oil trace element composition similar to the Libyan oil included in Table 5.4. The elemental concentration corresponding to a peak area of the nuclide used equal to 3 standard deviations of the background under the peak was taken as the detection limit for that matrix. The calculated results are shown in Table 5.6. It should be stressed that the actual detection limit for a given element in an oil may be higher or lower than the values in Table 5.6 depending on the contents of other trace elements.

Sources of Error

Fast neutron reactions are a possible source of error in the results. For example, ^{60}Co, used to determine Co, may be produced by the fast neutron reactions

$$^{60}Ni(n,p)^{60}Co$$

$$^{63}Cu(n,\alpha)^{60}Co$$

Shah, *et al*[35, 36] have evaluated these interfering nuclear reactions and have shown that the most serious interference is the ^{60}Ni(n,p)^{60}Co reaction,

Table 5.4

Trace Element Contents of Some Crude Oils
by Neutron Activation Analysis

Element	Concentration in Crude Oil[a]			
	California (Tertiary)	*Libya*	*Venezuela (Boscan)*	*Alberta (Cretaceous)*
V	7.5	8.2	1100	0.682
Cl	1.47	1.81	--	25.5
I	--	--	--	1.36
S	9.90	4694	--	1450
Na	13.2	13.0	20.3	2.92
K	--	4.93	--	--
Mn	1.20	0.79	0.21	0.048
Cu	0.93	0.19	0.21	--
Ga	0.30	0.01	--	--
As	0.655	0.077	0.284	0.0024
Br	0.29	1.33	--	0.072
Mo	--	--	7.85	--
Cr	0.640	0.0023	0.430	--
Fe	68.9	4.94	4.77	0.696
Hg	23.1	--	0.027	0.084
Se	0.364	1.10	0.369	0.0094
Sb	0.056	0.055	0.303	--
Ni	98.4	49.1	117	0.609
Co	13.5	0.032	0.178	0.0027
Zn	9.76	62.9	0.692	0.670
Sc[a]	8.8	0.282	4.4	--
U	--	0.015	--	--

a   Values in µg/g, except Sc, ng/g.

Table 5.5

Cadmium Contents of Nine Crude Oils and Two Asphaltenes

Sample	Location	Age	Cd Content (ng/g)[a]	
			Determination 1	Determination 2
Crude	California	Tertiary	4.09 ± 0.33	3.83 ± 0.36
Crude	California	Tertiary	<0.5	<0.5
Crude	Nigeria	—	1.22 ± 0.33	1.77 ± 0.41
Crude	Sumatra	Miocene	17.5 ± 0.51	16.8 ± 0.61
Crude	Alaska	—	0.32 ± 0.09	0.20 ± 0.11
Crude	Louisiana	U. Cretaceous	0.39 ± 0.07	0.35 ± 0.06
Crude	Utah	Eocene	0.60 ± 0.10	0.50 ± 0.11
Crude	Alberta	U. Devonian	9.6 ± 0.82	9.9 ± 0.91
Asphaltene	Alberta	U. Devonian	208 ± 22	241 ± 31
Crude	Alberta	U. Devonian	25.2 ± 0.49	29.1 ± 0.70
Asphaltene	Alberta	U. Devonian	1700 ± 300	1520 ± 290

a Error limits are standard deviations based on counting statistics.

Table 5.6

Detection Limits for Different Elements in a
"Typical" Crude Oil Matrix (in ng/g)

Element	Detection Limit	Element	Detection Limit
Cl	10	Ni	30
S	1000	Co	0.02
V	2	Se	23
Na	20	Hg	4*
As	6	Zn	90
Ca	4000	Cr	23
Mn	15	Fe	400
Cu	100	Sb	1.0
Ga	25	Sc	0.1
Br	1.5	U	100
I	5	Mo	500

*assumes Se absent.

and that even for a Ni/Co ratio in the oil of 570,
the error in the Co determination is only 6.7%. Fast
neutron interferences are thus negligible except for
Co in oils of high Ni/Co ratios. All results for
Co in Table 5.4 have been corrected for Ni interference.

Another possible source of error is loss of volatile compounds during irradiation due to radiolysis
of the petroleum matrix in a high γ-radiation field.
It was found that loss of light hydrocarbons occurred
during irradiation but examination of the gases
evolved showed no traces of Cl, Br, I, Hg, Ni, or
other trace elements measured in this study.

Precision and Accuracy

The precision of the neutron activation method
for 18 elements can be seen from the data in Table
5.7. The standard deviations listed have been calculated from replicate determinations of these elements
in a California oil. Except for Au, the relative
standard deviation for all elements is less than 12%.
The value of 57.1% for the relative standard deviation of the Au concentration is due possibly to
adsorption of ^{198}Au on the container walls during
irradiation because ^{198}Au was observed in some of
the irradiation vials after removal of the sample.
For this reason, Au contents of crude oils should
be interpreted with caution.

No well-characterized natural petroleum standards
calibrated for trace metals exist and thus it is difficult to evaluate the accuracy of the method.
As a check on the method, however, the National
Bureau of Standards Standard Reference Material SRM
1571 Orchard Leaves is analyzed with each batch of
crude oils analyzed. The results obtained for the
Orchard Leaves analyzed with a particular batch of
oils are shown in Table 5.8. The results obtained
for the elements listed in Table 5.8 show excellent
agreement with the National Bureau of Standards
Certified Values,[42] except for As. The NBS value
of 14 µg/g for As is currently being reevaluated and
is expected to be closer to 10 µg/g.

In an attempt to provide a trace element standard
for use in petroleum analysis the National Bureau
of Standards and the Environmental Protection Agency
recently distributed a Residual Fuel Oil (and other
materials) for analysis to more than 40 laboratories.
The results obtained and a discussion of the results
has been presented by La Fleur and Van Lehmden.[43]
The data present a disturbing picture of the state
of trace analysis of petroleum. Tables 5.9 and 5.10
show the results of the interlaboratory comparison
of Ni and V in the Residual Fuel Oil (RFO). The
largest range (13.6-52.4 µg/g) of Ni values was obtained by atomic absorption. Smaller ranges were
shown by other methods.

The data for V are more disturbing and the
range for atomic absorption is 22.4-444 µg/g which
is clearly unsatisfactory. For both Ni and V the
neutron activation values are very good and the
results obtained in this study agree well with pro-
visional values established by the NBS. Table 5.11
shows results for As, Cd, Cr, Hg, Mn, Pb, Se and Zn.

Table 5.7

Replicate Determinations of 18 Trace Elements
in Crude Oil

Element	No. of Replicates	Concentration (µg/g) Mean	S.D.	Relative S.D. (%)
Ni	5	93.5	2.4	2.5
V	10	7.5	0.6	8.0
Co	5	12.7	0.26	2.1
Hg	5	21.2	0.36	1.7
Fe	5	73.1	5.5	7.5
Zn	4	9.32	0.48	5.2
Cr	5	0.634	0.033	5.2
Mn	10	2.54	0.22	8.7
As	5	0.656	0.074	11.3
Au	5	2.8×10^{-6}	1.6×10^{-6}	57.1
Sb	5	0.0517	0.0011	2.1
Se	5	0.364	0.038	10.4
Sc	4	8.8×10^{-3}	6.7×10^{-4}	7.6
Cu	5	0.93	0.08	8.6
Na	5	11.1	0.33	3.0
Ca	10	192.0	14.0	7.3
Ga	9	0.346	0.006	1.7
Cl	9	2.35	0.16	6.8

Table 5.8

*Comparison of Trace Element Results Obtained
for SRM 1571 Orchard Leaves*

Element[a]	Concentration Determined	NBS Certified[b] Value[43]	Other Values[44]
Ni	1.6	1.3	
Fe	278	295	
Zn	27.5	25	
Co	0.160	0.16	
As	9.83	14*	10
Na	80.0	82	
Rb	11.4	12	
Br	11.1	(10)	
Cr	2.76	(2.3)	
Se*	81.7	80	
Cs*	61.0	--	<60
Sc*	80.0	--	200
Cd*	115	110	

a   Elemental concentrations in µg/g except for * concentrations in ng/g

b   Values in parentheses not certified

*   Value to be revised to 10 µg/g

Table 5.9

Interlaboratory Study of Ni in Residual Fuel Oil

Element	Method	No. Labs	Range (ppm)	Mean (ppm)
Ni	AAS	9	13.6–52.4	34.7
Ni	NAA	4	30.8–39.5	34.9
Ni	OES	3	34.5–60.4	44.1
Ni	XRF	5	22.0–40.4	32.7
Ni	other	2	35.5–39.0	37.3

NBS Certified Value: 37 ± 3 ppm

WSU Results: 37.4 ± 1.5 ppm

Table 5.10

Interlaboratory Determinations of V in Residual
Fuel Oil

Element	Method	No. Labs	Range (ppm)	Mean (ppm)
V	AAS	8	22.4–444	264
V	NAA	5	261–312	284
V	OES	2	150–299	225
V	XRF	5	230–313	277
V	other	2	287–310	299

NBS Certified Value: (315 ± 5) ppm

WSU Results: 312 ± 16 ppm

Table 5.11

Interlaboratory Determinations of Trace Elements
in Residual Fuel Oil

Element	Method	No. Labs	Range (ppm)	Mean (ppm)	Certified* Value
As	NAA	3	0.029-0.057	0.04	0.06
Cd	AAS	3	0.199-2.34	0.93	--
Cd	OES	2	2.72 -6.60	4.66	--
Cr	AAS	3	0.037-1.00	0.59	--
Cr	NAA	2	0.092-0.123	0.11	--
Cr	OES	2	0.079-0.197	0.14	--
Hg	AAS	6	0.021-0.399	0.15	0.002
Mn	AAS	5	0.145-0.305	0.27	0.121
Mn	NAA	2	0.120-0.251	0.19	0.121
Mn	OES	2	0.059-0.133	0.10	0.121
Pb	AAS	3	0.249-3.56	1.51	0.041
Se	NAA	2	0.172-0.191	0.18	0.138
Zn	NAA	2	0.320-0.483	0.40	0.17

*Probable certified or information values by NBS

Although the number of laboratories reporting is
small, the ranges of values for each element show
that there is little agreement among methods or
laboratories. The neutron activation values show
the smallest ranges and agree well with probable
certified values except for Zn. Results for 14
elements obtained in this laboratory by the neutron
activation method are shown in Table 5.12.

Table 5.12

Average Elemental Concentration in "Fuel Oil"[a]
(values are in ppb)

No.	Element	Mean	Standard Deviation	Number of Determinations
1	Chromium	92.6	9.60	9
2	Antimony	9.52	0.837	9
3	Cobalt	301.0	14.0	9
4	Selenium	172.0	18.4	9
5	Bromine	39.1	5.3	9
6	Nickel[b]	37.4	1.50	9
7	Zinc	320.0	30.0	9
8	Iron[b]	23.9	1.08	7
9	Sodium[b]	11.2	0.654	9
10	Arsenic	57.8	4.51	8
11	Mercury	<12		9
12	Manganese	252.0	9.49	8
13	Chlorine[b]	7.80	0.506	8
14	Vanadium[b]	312.0	16.4	8

a   Mean and standard deviation values are given to three significant figures.

b   These values are in ppm.

CONCLUSIONS

The neutron activation method for the determination of up to 25 elements in crude oils provides an
accurate and precise method for use in geochemical
studies of trace elements in petroleum.   The method
suffers from few interferences, and results obtained
with an NBS Standard Reference Material agree well
with certified values.   Methods of trace element
analysis of crude oils need to be evaluated for accuracy and precision because there are serious

discrepancies in the results obtained by different laboratories for several trace elements in a proposed trace element petroleum standard.

REFERENCES

1. Gordon, G. E. Private communication.
2. Milner, O. I. *Analysis of Petroleum for Trace Elements.* (New York: Macmillan, 1963).
3. McCoy, J. W. *The Inorganic Analysis of Petroleum.* (New York: Chemical Publishing Co., 1962).
4. Bonham, L. D. *Bull. Amer. Assoc. Petrol. Geol.*, *40*, 897 (1956).
5. Erickson, R. L., A. T. Myers and C. A. Horr. *Bull. Amer. Assoc. Petrol. Geol.*, *38*, 2200 (1954).
6. Hyden, H. J. *U. S. Geol. Survey Bull.* 1100-B (1961).
7. Hodgson, G. W. *Bull. Amer. Assoc. Petrol. Geol.*, *38*, 2537 (1954).
8. Katchenkov, S. M. and E. I. Flegentova. *Vestsi. Akad. Belarus. SSR Ser. Khim. Nauk. 95* (1970).
9. Chu-Kang, C. C., E. W. Kell, and E. Solomon. *Anal. Chem.*, *32*, 221 (1960).
10. Sinko, I., S. Gomiscek and M. Span. *Nafta* (Zagreb), *23*, 115 (1972).
11. Sandell, E. B. *Colorimetric Determination of Trace Metals*, 3rd ed. (New York: Interscience, 1959).
12. Smith, A. J., J. O. Rice, W. C. Shaner and C. C. Cerato. *Preprints Amer. Chem. Soc. Div. Petrol. Chem.*, *18*, 609 (1973).
13. Dickson, F. E., C. J. Kunesh, E. L. McGinnis and L. Petrakis. *Anal. Chem.*, *44*, 978 (1972).
14. Billings, G. K. and P. C. Ragland. *Can. Spectrosc.*, *14*, 8 (1969).
15. Gomiscek, S., M. Span and J. Sinko. *Nafta* (Zagreb), *23*, 29 (1972).
16. Alder, J. F. and T. S. West. *Anal. Chim. Acta.*, *61*,132 (1972).
17. G. Hall, M. P. Bratzel and C. L. Chakrabarti. *Talanta*, *20*, 755 (1973).
18. Filby, R. H., K. R. Shah and W. A. Haller. *J. Radioanal. Chem.*, *5*, 277 (1970).
19. Gordon, G. E., K. Randle, G. G. Goles, J. B. Corliss, H. M. Beeson and S. S. Oxley. *Geochim. Cosmochim. Acta*, *32*, 369 (1968).
20. Rey, P., H. Wakita and R. A. Schmitt. *Anal. Chim. Acta*, *51*, 163 (1970).
21. Guinn, V. P., R. A. Johnson and G. C. Mull. Sixth World. Cong. Proc. Sec. V Paper 20 (1963).
22. Jester, W. A. and E. E. Klaus. *Nucl. Appl. 3*, 37 (1967).

23. Braier, H. A. and W. E. Mott. *Nucl. Appl.* 2, 44 (1966).
24. Guinn, V. P. and S. C. Bellanca. NBS Spec. Publ. 312,
    Vol. 1, 93 (1969).
25. Patek, P. and H. Sorantin. *Allg. Prakt. Chem.* 24, 87 (1973).
26. Ordogh, M., A. Balasz, S. Reti and E. Szabo. *Banyasz.
    Kohasz. Lapok. Koolaj. Foldgaz.*, 6, 86 (1973).
27. Palmai, G. *Banyasz. Kohasz. Lapok. Koolaj. Foldgaz.*, 5,
    161 (1972).
28. Babaev, A. and M. U. Umarov. *Doklad. Akad. Nauk. Tadzh.
    SSSR 11*, 33 (1968).
29. Berkuptva, I. D., I. M. Zlotova, S. A. Punanova and
    K. I. Yakubson. *Neftegazov. Geol. Geofiz. 1970*, 19 (1970).
30. Lobanov, E. M., M. U. Umarov, I. M. Numanov and N. A.
    Azizov. *Doklad. Akad. Nauk. Tadzh. SSR 12*, 33 (1969).
31. *Doklad. Akad. Nauk. Tadzh. SSR 12* (6), 33 (1969).
32. Mast, R. F., R. R. Ruch and E. Atherton. *Ill. Geol.
    Survey Petrol.*, *No 95*, 111 (1971).
33. Arroyo, A. and D. Brune. *Mikrochim. Acta 1972*, 239 (1972).
34. Bryan, E. D., V. P. Guinn, R. P. Hackleman and H. R. Lukens.
    Gulf General Atomic Report GA-9889 UC-2 (1970).
35. Shah, K. R., R. H. Filby and W. A. Haller. *J. Radioanal.
    Chem.*, 6, 185 (1970).
36. *J. Radioanal. Chem.*, 6, 413 (1970).
37. Filby, R. H. and K. R. Shah. Proc. Amer. Nucl. Topical
    Meeting Nucl. Methods in Environment, Columbia, Mo.,
    Aug. 23-24, 1971. Univ. Missouri Press, p. 86 (1971).
38. Filby, R. H. Unpublished data.
39. Hitchon, B., R. H. Filby and K. R. Shah. Unpublished
    data.
40. Shah, K. R. and R. H. Filby. Internal Report, 1971.
41. Filby, R. H., A. I. Davis, K. R. Shah and W. A. Haller.
    Mikrochim. Acta, *1970*, 1130 (1970).
42. National Bureau of Standards Certificate of Analysis SRM
    1571 Orchard Leaves, Washington, D. C., 1971.
43. La Fleur, P. D. and D. Von Lehmden. Material distributed
    at Symposium on Trace Analysis of Coal, Fly Ash, Fuel Oil
    and Gasoline, Research Triangle Park, N.C., May 16-17,
    1973.

CHAPTER 6

GEOCHEMISTRY OF TRACE ELEMENTS IN CRUDE OILS, ALBERTA, CANADA

Brian Hitchon
Alberta Research
Edmonton, Alberta, Canada

and

R. H. Filby and K. R. Shah
Nuclear Radiation Center
Washington State University
Pullman, Washington

INTRODUCTION

This chapter reports preliminary results from
a comprehensive, on-going study of the relations
between trace elements and the organic components,
specifically asphaltenes, in a large suite of crude
oils and oil sands from Alberta, Canada. We report
here only the results of the initial examination of
the interrelations among 22 trace elements in a
limited sample of 88 crude oils from stratigraphic
units ranging in age from the Upper Cretaceous,
Belly River Formation, to the Middle Devonian, Elk
Point Group, and underlying Granite Wash. They are
representative of all major petroleum-producing
stratigraphic units in Alberta and range geographi-
cally throughout the province. In type, they include
condensates, crude oils, and two samples of the heavy
oil from the McMurray Formation at the Athabasca
deposit. Table 6.1 shows the number of samples
examined from specific stratigraphic units.
The crude oils were fresh samples from cold
separators, or cold treaters, or from a cold tank
recently produced into from a cold separator or cold
treater, and without the addition of emulsion break-
ers. As far as is known, downhole corrosion inhibitors

111

Table 6.1

Stratigraphic Subdivisions of 88 Crude Oils
Analyzed for Trace Elements

Stratigraphic Unit	Number of Samples
Upper Cretaceous, Belly River Fm.	4
Upper Cretaceous, Upper Colorado Gp.	7
Lower Cretaceous, Viking Fm.	10
Lower Cretaceous, Mannville Fm. (including oil sands)	19
Jurassic	3
Triassic	4
Carboniferous	8
Upper Devonian, Wabamun Gp.	2
Upper Devonian, Winterburn Gp.	6
Upper Devonian, Woodbend Gp.	15
Upper Devonian, Beaverhill Lake Fm.	4
Middle Devonian, Elk Point Gp.	5
Granite Wash	1

had not been added to any well sampled. The samples were collected in standard sealer jars with glass tops and O-rings. Preliminary clean-up included centrifuging for half an hour at 2500 rpm and filtration through Whatman No. 1 filter papers; the samples were then transferred to glass bottles with caps with teflon inserts. Further filtration through Millipore 5μ filters followed the sequence described by Filby.[1]

Trace elements were determined on the 88 crude oils by neutron activation analysis using the multi-element method of Shah, *et al.*[2,3] and Filby and Shah.[4] The general statistics for the results of these determinations are shown in Table 6.2. In a few instances it was not possible to determine

Table 6.2

General Statistics for 22 Elements Determined in
88 Crude Oils (Total Oil Basis), Alberta, Canada

Element	Unit of Measurement	Detection Limit	Number of Samples Above Detection Limits	Average Concentration	Maximum Concentration
S	%	0.05	88	0.83	3.88
V	ppm	0.1	84	13.6	177
Cl	ppm	–	87	39.3	1,010
Na	ppm	–	85	3.62	64.7
Fe	ppm	0.1	41	10.8	254
Ni	ppm	0.1	69	9.38	74.1
Zn	ppb	25	79	459	5,920
Co	ppb	0.2	84	53.7	2,000
I	ppb	10	51	719	9,000
Mn	ppb	3	78	100	3,850
Se	ppb	3	62	51.7	517
Hg	ppb	2	39	50.9	399
Cs	ppb	0.5	43	4.27	68.5
Br	ppb	2	82	491	12,500
As	ppb	2	71	111	1,990
Au	ppb	–	4	0.438	1.32
Sb	ppb	0.01	24	6.22	34.8
Cr	ppb	5	42	93.3	1,680
Rb	ppb	10	9	148	720
Sc	ppb	0.05	51	7.76	199
Eu	ppb	0.05	50	0.935	23.2
Ga	ppb	–	–	–	445

sulfur in some samples, and values from chemical
determinations were substituted in the statistical
study. Further, gallium was not determined on all
samples and so average concentrations have not been
given in Table 6.2.

The results of these determinations for 22 elements on 88 crude oils were subjected to factor
analysis, which is a statistical technique designed
to explain complex relations among many variables
in terms of a few factors which themselves represent
simpler relations among fewer variables. Factor
analysis only determines the relations; it does not
explain them. The explanation of the factors must
be in the context of known information about the
variables. Hitchon and Gawlak[5] have used factor
analysis in the study of aromatic hydrocarbons in gas
condensates from Alberta and their paper includes
pertinent background information with particular application to geochemical problems.

RESULTS OF FACTOR ANALYSIS

Q-Mode

Q-mode factor analysis consists of a comparison
of the samples in terms of the variables. This
technique essentially evaluates the homogeneity of
the sample or population being studied. Most elements
essentially exhibited a lognormal distribution and
so the logarithmic transforms of the geochemical data
were factored and not the raw data. It is not essential to carry out the transformation, although it
is desirable if sound conclusions are required based
on significance testing.[6] Where a large number of
determinations for a specific element were below
detection limits, that particular element could not
be used in the factor study. Further, to enable the
logarithmic transformation to be made, determinations
below detection limits were arbitrarily assigned
values one-tenth that of the respective detection
limit (Table 6.2). Accordingly, only eleven elements (S, V, Cl, Na, Ni, Zn, Co, Mn, Se, Br and As)
were used in the general factor analysis.

The Q-mode varimax factor score matrix of the
trace element data is shown in Table 6.3. The first
factor has high positive scores for Br, As, Zn, V,
Ni, Se, Co and Mn; the second factor has high positive scores for Ni, V and S opposed by high negative
scores for Mn, Zn and Br; and the third factor has

Table 6.3

*Q-Mode Varimax Factor Score Matrix of Trace
Element Data for 88 Alberta Crude Oils*

Variable	*Factor*		
	1	*2*	*3*
S	0.20	0.77	0.38
V	1.09	1.71	-0.25
Cl	0.69	-0.53	-0.52
Na	0.18	0.57	0.22
Ni	1.03	1.81	-0.31
Zn	1.24	-1.07	-1.87
Co	0.90	0.22	-0.08
Mn	0.83	-1.16	-0.37
Se	0.95	0.39	-0.95
Br	1.59	-0.88	1.01
As	1.30	-0.38	2.19
Percent of Information Explained by Factor			
	43.7	30.6	9.6
Cumulative Percent of Information			
	43.7	74.3	83.9

a very high positive score for As, a high positive
score for Br, and low negative scores for Zn and
Se. The varimax factor components for these three
factors may be normalized and the data points plotted on a ternary diagram (Figure 6.1). Only a few
selected series of points have been plotted to illustrate the main features. With only five exceptions, the data points fall in the region with a
less than 40% component of factor 3 (dashed line in
Figure 6.1), and essentially form a binary series
between factor 1 and factor 2. The arrows indicate
present hydrodynamic flow paths based on published
information from Hitchon[7] and include data from
the Viking Formation with its well-marked hydraulic

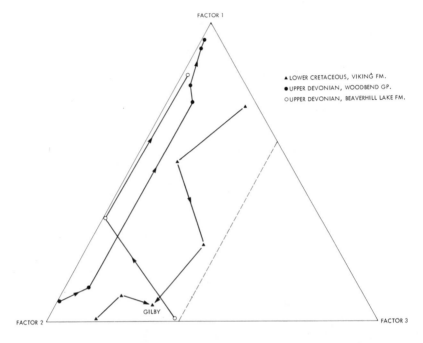

*Figure 6.1. Ternary diagram of normalized varimax factor
components for Q-mode on crude oils from
Alberta, Canada, showing relation of factor
loadings to hydrodynamic flow path for three
selected series of crude oils.*

head low at Gilby. Factor 1 represents the shal-
lower, less mature crude oils with high contents
of a wide variety of trace elements. Factor 2
includes the deeper, more mature crude oils in which
there are lower contents of trace elements but in
which Ni, V and S are still important, relative to
the other trace elements, even though in total
quantity they are generally lower than in the shal-
lower crude oils. This observation is true for all
stratigraphic units, including the Viking Formation
in which reverse, down-dip flow has been demonstrated.
It thus appears that the binary system represented
by factors 1 and 2 is essentially a maturation and
not a migration phenomenon. This implication must
be weighed in the context of variable amounts of the
asphaltene fractions of these 88 crude oils, which
range from essentially zero to about 17%,[8] and
taking into account the fact that a large fraction
of the trace elements is present in the asphaltene

fraction of crude oils, as observed by many workers.
With this caution in mind, it would appear that
the 88 crude oils represent a single population,
as far as the total trace elements are concerned,
which has been subjected to differentiation of the
trace elements through maturation, but not through
migration--at least as evidenced by the present
flow paths. Only detailed examination of the trace
elements in the asphaltene fraction will resolve
this problem. The result of the Q-mode analysis
implies that we are justified in examining all the
samples as one population by R-mode analysis.

R-Mode

     R-mode factor analyses using both the varimax
and biquartimin solutions were carried out on the
same data used for the Q-mode. In the varimax
analysis, the communalities were all greater than
0.8, indicating that most of the variance of each
component is explained by the seven factors extracted.
These seven factors were subjected to the oblique
biquartimin solution, and the loadings of the variables on the factors are illustrated in Figure 6.2.

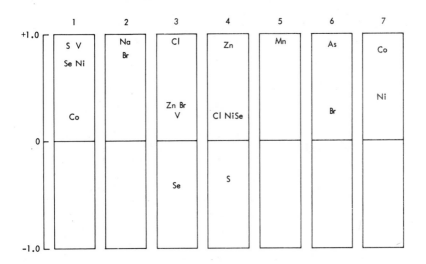

Figure 6.2. Diagrammatic representation of factors from
             R-mode biquartimin solution of trace element
             data for 88 crude oils from Alberta, Canada.

The rectangular boxes represent the factors, and
the center line of each box is a zero loading for
the component. Positive loadings (to a maximum
+1.0) occur above the center line and negative loadings (to a maximum -1.0) below. The further a component from the center line the higher is its loading.
Loadings less than 0.2 have been omitted from the
figure since they correspond, approximately, to less
than 5% of a variable. The seven factors account for
over 90% of the total variance among the variables.
A correlation matrix between the primary factors indicates that the factors are uncorrelated, with no
correlation coefficient exceeding 0.24.

More than one quarter of the cumulative variance
is accounted for by factor 1, which has high loadings
for S, V, Se and Ni. Essentially all the variance
of V and much of the variance of Ni is loaded onto
factor 1. This, together with the well documented
occurrence of vanadyl and nickel porphyrins in crude
oils and the fact that the major portion of the sulfur in crude oils occurs as organic compounds, suggests that factor 1 represents trace elements present
as metallo-organic compounds. The known close geochemical association of Se and S would account for
the high loading of Se on factor 1 and, by analogy,
imply that Se occurs in crude oils in the form of
Se-organic compounds. It is interesting that Na and
Cl are loaded on separate factors, since it indicates
that the crude oils have been separated successfully
from any entrained formation waters--which are essentially sodium chloride solutions in most cases in
Alberta. However, it leaves unresolved a reasonable
explanation for the loadings observed for factors 2
and 3.

Due to the large number of samples in which Fe
and Cr were below detection limits it was not possible to include them in the general R-mode analysis.
However, a smaller suite of 20 crude oils was selected
in which V, Fe, Ni, Co, Mn and Cr were all found
at concentrations above detection limits. A separate
R-mode analysis was run on these 20 samples and the
results of the biquartimin solution are shown in
Figure 6.3.   The four factors extracted account for
nearly 95% of the total variance among the variables.
The correlation coefficient between Fe and Co in 38
crude oils in which both elements were above detection
limits was 0.42, and the correlation coefficient between Fe and Mn in 36 crude oils in which both elements were above detection limits was 0.18. From
these various facts we can state, with some confidence, that Fe is unrelated to either Co or Mn in

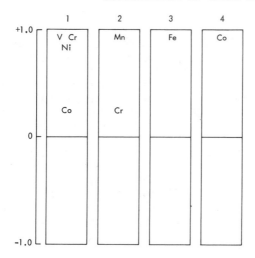

*Figure 6.3. Diagrammatic representation of factors from
R-mode biquartimin solution of 6 trace elements
(Fe, Co, Ni, V, Mn, Cr) for 20 selected crude
oils from Alberta, Canada.*

Alberta crude oils; nor is it related to V, Ni or Cr.
This suggests that Fe in the Alberta crude oils is
probably not present as iron porphyrins--which, as
far as we are aware have never been detected in crude
oils, although they have been reported from the
Colorado oil shale[9] and a variety of argillaceous
rocks.[10] Filby[1] has shown that Fe-porphyrins, if
present, account for a negligible fraction of the
Fe in a California Tertiary oil containing 68.9 ppm
Fe. The two orders of magnitude between the average
content of Fe and those of Mn and Co in the Alberta
crude oils (Table 6.2) might suggest that these
metals were present as a result of the incorporation
of corrosion products from well casing, tubing,
separators, treaters or storage tanks. We believe
that the demonstrated relations revealed by the
factor analyses and correlation coefficients effec-
tively negate this supposition. In oils studied
from other parts of the world, Filby[1] also found
much higher Fe contents than Mn or Co. However,
again, it leaves unresolved a reasonable explanation
for the loadings for Fe, Mn and Co in the two factor
studies.

Factor 6 represents an essentially unique factor
for As. In factor 4, the small opposed loading of S
against Zn could be interpreted as preferential complexing of Zn by N- or O-containing ligands.

This preliminary factor analysis leaves many questions unanswered concerning the relations among the trace elements in the total crude oils. But, as pointed out previously, the fact that most trace elements are associated with the asphaltene fraction of crude oils, together with the facts that the contents of asphaltenes vary widely and hence that many elements were below detection limits, indicates that examination of trace elements in the asphaltenes should resolve many of the uncertainties posed by the present study. Furthermore, EPR, NMR and elemental (C, H, N, O, S) analysis of the asphaltene fraction may help link the factors controlling the organic and inorganic components in these Alberta crude oils.

CONCLUSIONS

Within the limitations imposed by the determination of trace elements in the total crude oil, when their concentration is essentially confined to a variable portion of the total crude oil, we may conclude that the contents of the trace elements S, V, Cl, Na, Ni, Zn, Co, Mn, Se, Br and As are controlled by maturation processes rather than migration processes, though this may simply reflect maturation of the asphaltene fraction of the crude oils. S, V, Se and Ni account for more than one quarter of the cumulative variance and represent metallo-organic compounds. The factor analyses further demonstrate not only the efficacy of our separation systems between crude oil and entrained formation water, but show that it is most unlikely that the Fe present in the crude oils originates from corrosion products incorporated during production. Four elements, Fe, Mn, As and Co represent essentially unique factors. Finally, it is clear that considerably more work is justified on trace elements in crude oils, particularly in relation to their distribution within the various fractions and in relation to interactions between organic and inorganic components.

REFERENCES

1. Filby, R. H. ACS Symposium "Role of Trace Metals in Petroleum," Chicago (August 26-31, 1973).
2. Shah, K. R., R. H. Filby and W. A. Haller. *J. Radioanal. Chem.*, *6*, 185, (1970).
3. Shah, K. R., R. H. Filby and W. A. Haller. *J. Radioanal. Chem.*, *6*, 413, (1970).
4. Filby, R. H. and K. R. Shah. Proc. Amer. Nucl. Soc. Topical Conference on Nuclear Methods in Environmental Research, University of Missouri, Columbia, Mo. (August 23-24, 1971). p. 86.
5. Hitchon, B. and M. Gawlak. *Geochim. et Cosmochim. Acta*, *36*, 1043, (1972).
6. Harmon, H. H. *Modern Factor Analysis* (Chicago: Univ. of Chicago Press, 1967), p. 374.
7. Hitchon, B. *Water Resources Research*, *5*, 460, (1969).
8. Speight, J. G. Alberta Research, personal communication.
9. Moore, J. W. and H. N. Dunning. *Ind. Eng. Chem.*, *47*, 1440, (1955).
10. Hodgson, G. W., B. Hitchon, K. Taguchi, B. L. Baker and E. Peake. *Geochim. et Cosmochim. Acta*, *32*, 737 (1968).

Contribution No. 693 from Alberta Research.

CHAPTER 7

METALS IN NEW AND USED PETROLEUM PRODUCTS
AND BY-PRODUCTS
QUANTITIES AND CONSEQUENCES

Ivan C. Smith, Thomas L. Ferguson
and Bonnie L. Carson
Midwest Research Institute
425 Volker Boulevard
Kansas City, Missouri 64110

PHYSIOLOGICAL EFFECTS AND
MODES OF TRANSPORT OF METALS

A more accurate definition of the role of metals
in human health is needed if we are to relate health
or physical-disorder patterns to the human intake
of metals. Almost all the stable metals are found
in living matter. Fossil fuels contain many of these
metals, some in relatively large quantities. Petroleum products, which supply about 43% of the nation's
energy requirement,[1] are therefore a potential source
of man-made environmental contamination by metals.

Trace metals can be divided into three categories:
those known to be dietary essentials, those possibly
essential, and nonessentials. Metals known to be
essential to higher animals include: chromium, cobalt,
copper, iron, manganese, molybdenum, selenium, tin,
vanadium, and zinc. Several additional metals considered as possibly essential on the basis of sugges-
tive but inconclusive evidence include: arsenic,
barium, cadmium, nickel, and strontium. The other
metals are presently classified as nonessential.

Many of the trace metals occur in animals in
quantities that reflect the contact of the animal with
its environment. It has been suggested that the
essential elements can be distinguished from nonessential elements by observing the distribution patterns

123

in tissue. Some researchers have postulated that
internal mechanisms control the levels of essential
trace metals in tissue, whereas nonessential metals
occur in distribution patterns that reflect the
environment.[2]
    Some trace metals are classified as toxic. There
is, perhaps, justification for this classification
for such metals as arsenic, lead, and mercury. In
addition, extended exposure of mammals to small amounts
of cadmium, lead, selenium, antimony, and nickel carbonyl can shorten life or cause cancer, and lead,
nickel, antimony, cadmium, and mercury in small amounts
cause human health problems.[3] However, all metals
are toxic if ingested at sufficiently high levels.
Frequently, the effects of a toxic metal are increased
by nutritional deficiencies.[4]
    It is also becoming increasingly apparent that
the specific chemical form of a metal is a factor
in its toxicity. Water-lipid partitioning, cell-wall
permeability, and metabolic transport are just a
few of the factors influenced by chemical form which,
in turn, influence toxicity. Hexavalent chromium,
pentavalent vanadium, divalent manganese, arsine,
and methylmercury are more toxic than other forms
of the corresponding metal.[4]
    Synergistic effects such as those of cadmium
and cyanide or of lead and chromium are just beginning
to be recognized.[5] Antagonistic effects such as those
between selenium and arsenic, fluoride, mercury, or
lead; cadmium and zinc or selenium; copper and zinc
or molybdenum; and iron and manganese or zinc have only
recently been observed.[5-8]
    Table 7.1 summarizes some of the known physio-
logical effects of metals that are found in crude
oils, refinery products, and used petroleum products.
In appraising the potential environmental hazards of
metals found in petroleum, it is necessary to consider
where they enter the environment, how they move in
the environment, and how they might enter the food
chain.
    Metals applied to land are relatively immobile,
particularly if the soil has a high pH and high organic content. Many complex physical-chemical processes affect metal mobility in soil. Lead and mercury,
which form strong complexes, move very slowly. Zinc,
cadmium, and copper move a little faster. Metal
profiles in highly contaminated soil penetrate little
below the 4-8 inch region even after 60 years.[38]
    Accumulation of metals by plants grown on these
highly contaminated soils depends on plant variety
and soil conditions. In most cases, plants will not

concentrate sufficient metal to pose a serious human
or animal health problem. There are some exceptions,
such as selenium. Often the metal uptake by plants
can be reduced by proper soil management.
Metals introduced into surface waters again pose
an unknown hazard. Water-quality criteria have been
established to protect the populace (Table 7.2),
but in many cases these standards are based on inade-
quate scientific data.
Most metals disposed of in surface water are
rapidly removed from the aqueous phase at a rate
largely dictated by the amount of sediment or suspended
solids and eventually settle out. Once trapped in the
sediment, they are relatively immobile; however, there
will always be some freedom of movement of the metals
between the sediment and the overlying water. Aquatic
plants such as algae are excellent scavengers and
concentrators of metals in surface waters. These
plants are important to the lower end of the food
chain, and serve as a source of metals; they can be
passed up the food-chain ladder to the point where
they ultimately serve man as a food. Contamination
of groundwater is usually not a problem unless surface water enters the groundwater supply directly,
e.g., through a surface fault.
Metals lost to the environment as air pollutants
pose a direct health hazard if inhaled or ingested;
otherwise, they eventually enter soil or water and
their mode of transport is dictated by these media.
We know that metals lost to the environment will not
degrade. We also know they affect plant, animal,
and human health; yet we are just beginning to relate
environmental trace metals levels to health and disease
patterns.
Relatively few recognized incidents of human-
health problems have been attributed to metal contamination of the environment. Consumption of rice grown
in cadmium-contaminated irrigation water is generally
assumed to be the cause of the Japanese Itai-itai
disease, but other factors such as nutritional deficiencies may have been important.[4] Cadmium dissolved
from galvanized-iron water pipes in soft-water areas
has been linked to a higher incidence of hypertension
and heart attacks.[3] The Minamata Bay incident in
Japan was caused by environmental organomercury contamination. Natural selenium contents of some alkaline
soils are associated with many disease symptoms in
humans and livestock.[40] Vanadium air pollution is
implicated by statistical studies in deaths from
heart disease, hypertension, and digestive-system
cancer.[3] Nickel contamination is of concern because
of its possible role in the higher incidence of lung
cancer in urban populations.[4]

Table 7.1

Physiological Effects of Metals Found in Petroleum Products

Metal	Human Effects	Animal Effects	Plant Effects	Toxicity Data
Antimony	Dermatitis, keratitis, conjunctivitis, and nasal septum ulceration by contact fumes or dust.[9]	Shortens lifespan when fed to rats and mice.[10] Sb oxide caused pneumonitis and heart and liver damage.[11]	--	>0.1 g lethal oral dose in humans.[12]
Arsenic	Dermatitis, bronchitis, skin cancer.[13] Damages the heart,[14] kidney,[15] nerves,[15] and possibly the liver.[16] GI symptoms in acute systemic poisoning.[15]	Counteracts the toxic effects of Se in rats and chickens.[17] As(V) is nontoxic.[18] Morphol. changes in blood; kidney damage.[16]	0.1 ppm AsO₂ reduces heterotrophic activity of freshwater micro-flora.[19]	Normal ingestion 0.1 mg/day. Toxic level 5-50 mg/day.[20] Smallest fatal dose recorded 130 mg.[12]
Barium	Baritosis, a benign pneumoconiosis.[21] Soluble salts highly toxic orally.[12] Soluble salts are skin and mucous membrane irritants.[15]	Full strength Ba lubricant dispersant is a mild eye irritant.[22] BaO and BaCO₃ caused paralysis.[11]	Poisonous to most plants.[12] 0.1 ppm Ba²⁺ reduces heterotrophic activity of freshwater micro-flora.[19]	LD₅₀ diesel exhaust solids >10 g/kg (animals).[21] LD₅₀ (oral) Ba sulfonates, 3-10 g/kg.[22] LD₅₀ (oral) Ba phenolates 4-5 g/kg.[22] Fatal dose in humans >0.55 g BaCl₂.[23]

Element			
Boron	Therapeutic use of H_3BO_3 and borax has caused fatalities. CNS depressant and GI irritant. Boranes highly toxic.[12] Cumulative poison.[11]	2,500 mg/liter in drinking water inhibited animal growth.[16]	0.5-1.0 ppm in soil required for growth of fruit trees. 2.0 ppm possibly toxic.[24] Prevents pitting of apples.[3] Most B fertilizer used on alfalfa.[25]	H_3BO_3 fatal dose in adults 15-20 g and in infants 5-6 g.[11]
Cadmium	Cumulative poison.[12] Pulmonary emphesema, hypertension, kidney damage.[13] Cardiovascular disease.[26] Interferes with Zn and Cu metabolism. Inhalation of 0.03-35 mg/m^3 significantly reduced children's weight.[16] GI symptoms.[11]	1 ppm present in many plant and animal tissues.[12] Caries, anemia, and retarded growth in rat.[17] Maximum concentration in earthworms near a highway, 115 ppm.[27]	Stunts growth of lettuce, radish,[28] bean and turnip plants. Tomatoes, barley and cabbage more tolerant. Leaves accumulate excessive amounts when solutions contain a few tenths µg/ml.[29]	LD (oral) for rabbits 200-600 mg. LD_{50} CdO fume 500 mg/m^3 for rats to 1,500 mg/m^3 for monkeys.[12]
Chromium	Dermatitis,[15] ulceration of skin,[11] perforation of nasal septum, chronic catarrh, emphysema, carcinogenesis when inhaled.[13,23] Cr(VI)[30] extremely toxic. Not cumulative. Apparently essential in glucose metabolism.[17]	100 ppm recommended limit for fisheries.[30]	5 ppm recommended limit for irrigation water.[30] Soils contg. 0.2-0.4% Cr are infertile.[12] Toxic to some aerobic microbes at ppb level. A micronutrient.[31]	Normal ingestion 0.05 mg/day. Toxic level 200 mg/day.[20] No ill effects from well water with 1.0-25.0 mg/liter.[16] LD (oral) K chromate in rabbits 1.9 g within 2 hr.[12]

Table 7.1, continued

Metal	Human Effects	Animal Effects	Plant Effects	Toxicity Data
Cobalt	Goitrogenic. Lung effects disputed. Dermatitis. No injury from $Co_2(CO)_8$.[12] Affects heart and GI tract.[14] $\sim 7 \mu g$/day beneficial.[23] Liver and kidney damage.[29]	Essential nutrient. Polycythemia. Bone hyperplasia. Metaplasia in spleen, liver and kidneys. Hyperglycemia due to reversible pancreatic damage.[12]	Essentiality not established.[24]	Normal ingestion 0.002 mg/day; toxic level 500 mg/day.[20] Co metal dust more toxic than salts in lung irritation; lethal dose of either relatively high.[12] LD $Co_2(CO)_8$ in animals by inhalation 100 ppm.[12]
Copper	Antagonistic to Zn toxicity.[7] Not cumulative. Require 1-2 mg/day.[12] Inhalation of Cu-contg. dust causes lung and GI disturbances.[12] Affects erythrocytes and liver.[14] Skin and mucous membrane irritants.[15]	Ingestion or inhalation caused hemochromatosis and lung and liver injury.[12] Limits: irrigation water, salt-water organisms, and freshwater organisms, 0.1, 0.05, and 0.02 ppm, respectively.[30]	A micronutrient.[31] Toxic to some aerobic microbes at ppb level.[31] 0.05 M Cu^{2+} inhibits root growth and 0.1 M stops germination of lettuce.[32]	Normal ingestion 2-5 mg/day; 65-130 mg $CuSO_4$ dangerous and 648-972 mg highly toxic.[12] 27 g $CuSO_4$ fatal.[11]

Iron	Siderosis (a pneumoconiosis due to Fe inhalation).[13]	—	A micronutrient.[31]	—
Lead	Brain damage, convulsions, behavioral disorders, death.[26]	Pb naphthenate kills rabbits by skin absorption (death due to pneumonia).[22] Cumulative poison in vertebrates.[30] Renal and vascular poison.[12]	0.1 ppm Pb^{2+} reduces heterotrophic activity in microflora.[19]	Oral toxicity of Pb naphthenate 3.5-5.1 g/kg.[22] Normal Pb ingestion 0.4 mg/day.[12]
Manganese	Chronic Mn poisoning and/or Mn pneumonitis.[13] Reduces Fe absorption.[14] Requirement 3-9 mg/day.[2] Primarily a nerve toxin.[12] CNS symptoms often result in permanent disability.[15]	Pathological effects on nerve cells and the liver.[12] 1.9-9.9 mg is thyroid inhibitor in rats.[16]	A key role in photosynthesis.[31]	
Mercury	Nerve damage and death.[26]	Detrimental to aquatic ecosystems at 0.005 ppm.[30]	0.1 ppm Hg^{2+} reduces heterotrophic activity of microflora.[19]	Normal ingestion 0.005-0.2 mg/day; toxic level 10 mg/day.[12]

Table 7.1, continued

Metal	Human Effects	Animal Effects	Plant Effects	Toxicity Data
Molybdenum	No indication of even[12] an industrial hazard.[11] Not cumulative.[11]	Appears to have a reciprocal antagonism with Cu. Requirement in rats <0.5 mg/day.[12] Low order of toxicity.[15] No fatalities from molybdic oxide fumes for 25 one-hour exposures at 1.5 mg/ft³ air.[11] Toxic to ruminants when fed in excess.[33]	Essential to higher plants.[12] Possible role in photosynthesis.[31]	Ingestion of <500 mg/day MoS2 nontoxic to animals. 8.1 mg MoS2/ft³ nontoxic to guinea pigs. MoO3 at 5.8 mg/ft³ very irritating with high mortality. MoO3 dust more toxic than fume.[12]
Nickel	Rarely gives systemic toxic effects even from therapeutic doses (65-195 mg NiSO4, and 324-454 mg NiBr2). Dermatitis, respiratory disorder, carcinogenesis (nose and lung).[13]	Moderately toxic to aquatic organisms.[30] Maximum Ni concentration in earthworms near a highway 38 ppm.[27] Inhibits enzyme systems. Kidney damage (calf).[17] 1-3 mg/kg Ni compound causes intestinal disorders, convulsions, and asphyxia in dogs.[11]	Can be very toxic depending on its chemical form.[30] A micronutrient.[31] 0.01 ppm reduces the heterotrophic activity of freshwater microflora.[19]	Normal ingestion 0.3-0.5 mg/day.[12] 30-73 mg NiSO4 6H2O toxic in humans.[23]

Selenium	May cause caries.[26] Prevents teratogenic effects of Cd and As.[14] Affects kidneys, liver, marrow, and CNS.[34] Se compounds are potent skin and mucous membrane irritants.[15]	Carcinogenic in large doses in rats.[26] Essential to mammals and chicks in low doses.[26] Teratogenic in chicks.[35] "Blind staggers" and "alkali disease" in cattle and "white muscle disease" in sheep.[12]	2.5 mg/liter Se inhibits the BOD and growth of aquatic saprophytic microflora.[34]	Industrial selenosis symptoms when Se in air <0.2 ppm.[34] Liver damage in humans from 5–7 mg/liter in food.[16] Liver cancer in animals from food containing 10 ppm Se.[36] H_2Se and SeO_2 more toxic than S analogs. LD by inhalation of SeO_2 10 ppm for 2 hr.[36] Normal ingestion 0.2 mg/day; toxic level 5 mg/day.[20] Extreme tolerance limit in food (dry weight) 20 ppm.[20]
Silver	Argyria (impregnation of the tissues with Ag) following absorption from the GI tract or lung. I.v. injections of colloidal Ag fatal.[12]	Affects immunological capacity. Histopathological changes in tissues of encephalon and medulla of rabbits.[16]	0.0001 ppm Ag^+ reduces the heterotrophic activity of microflora.[19]	

Table 7.1, continued

Metal	Human Effects	Animal Effects	Plant Effects	Toxicity Data
Tin	Little absorbed when ingested.[12] Affects GI tract and CNS.[14] Sn oxide dust produces benign pneumoconiosis.[37] Ingestion of organotin compounds causes acute cerebral edema and often death.[15]	Decreases longevity slightly when fed to rats and mice for life.[10]	Toxic to some aerobic microbes at ppb levels.[31]	500 mg/kg/day of $SnCl_2$ for 14 months paralyzed a dog.[16] LD_{50} (i.v.) R_2SnCl_2 (where $R = \leq C_8$ alkyl) 5-40 mg/kg rats. Trialkyltin salts are more toxic.[11]
Vana-dium	Cardiovascular disease, carcinogenesis.[13] Main toxic effects on respiratory system.[12] V_2O_5 residues from fuels irritating to those who clean oil-fired burners, renew firebrick linings, and clean heat-exchanger tubes (dusts contain 6-20% V).[12]	Feeding 5 mg/liter V^{4+} for life to rats gave no significant reduction in growth or longevity.[16] Not carcinogenic in rats and mice.[10]	Essential for green algae. Stimulates higher green plants in small amounts.[12]	10 mg/kg fatal to rat.[16] Sublethal doses 92-368 ppm. 49 μg/ml drinking water highly toxic. 0.205 mg/liter causes lung changes in animals. LD (i.v.) in humans 30 mg V_2O_5 as tetravanadate.[12]

| Zinc | Dermatitis, hypertension, arteriosclerotic and heart diseases.[13] 675-2,280 mg/liter is emetic.[23] Causes mineral loss from bones.[16] Most Zn compounds not particularly toxic at moderate concentrations orally.[12] Zn inhibits the teratogenic, embryocidal, and neo-plastic effects of Cd.[14] Essential. | Zn dithiophosphates are severe eye irritants.[22] 0.1-1.0 ppm lethal to fish and other aquatic animals.[30] | Essential. 0.1 ppm Zn^{2+} reduces heterotrophic activity of microflora.[19] | LD50 (oral) Zn dithiophosphates 2.13-3.7 g/kg. LD50 (skin) 11.3 g/kg for rabbits (24-hr contact). Mixed Mg-Zn phenolate LD50 (oral) 9.5 mg/kg.[22] LD50 (oral) $ZnCl_2$ in guinea pigs, rats, and mice 200-350 mg/kg. Normal ingestion Zn 10-15 mg/day.[12] |

Table 7.2

Surface Water Criteria for Public Water Supplies*

Metal	Permissible Criteria (mg/l)
Arsenic	0.05
Barium	1.0
Boron	1.0
Cadmium	0.01
Chromium (hexavalent)	0.05
Copper	1.0
Iron (filterable)	0.3
Lead	0.05
Manganese (filterable)	0.05
Selenium	0.01
Silver	0.05
Zinc	5.0

*from Reference 39.

Complete assessment of the effects of chronic human exposure to low concentrations of physiologically active metals cannot be made until all the synergistic and antagonistic relations between metals and between metals and other environmental factors have been identified and understood.

There are two routes to developing the informational needs on the role of trace metals in health and disease. The first is to conduct fundamental studies on the physiological effects of metals and metal combinations. Such studies are extremely expensive and time-consuming and would require an almost infinite number of combinations of metals and chemical forms of metals if this approach were used.

The second approach is to use epidemiological studies to relate health and disease patterns statistically to known environmental contamination. Although this approach is fraught with possible misinterpretations, it is a valuable tool for identifying the important relations between environmental contamination and health and disease. Such studies would, however, take many years to complete and would require the use of virtually all available analytical resources

to develop the needed data. Alternatively, a more
practical method for obtaining needed information on
environmental contamination is to identify sources
and quantities of metals entering the environment
and where they enter the environment.

METALS IN CRUDE PETROLEUM
AND PETROLEUM PRODUCTS

Efforts to develop standardized analytical procedure for measuring the metal content of petroleum
products have been only partially successful; the
literature is replete with inaccurate data. However,
even allowing for substantial errors, available data
indicate that petroleum products contain substantial
quantities of metals.
Crude oil contains varying amounts of metals,
depending on the source of the crude. As a general
rule, the heavier the crude, the more metal is
present. Table 7.3 identifies the elements found
in crude oil and their range in concentration. Using
the medians of the concentration ranges, one can calculate the total pounds of metal that may enter the
environment each year in the United States as a result
of crude-oil uses. The various products derived from
crude oil are shown in Table 7.4.
The ultimate fate of metals in crude petroleum
has never been determined. An assessment of available information indicates that the majority of the
metal in the crude oil is retained in the bottoms that
remain after distillation. A small amount of the
metals is carried over in the various distillate fractions. The ranges of concentrations of metals found
in five distillate fractions are shown in Table 7.5.
Data recently reported for gasoline show ppm concentrations of silver (0.01-0.1), cadmium (0.01-0.05),
copper (0.01-5), manganese (0.01-0.1), zinc (0.1-10),
and lead(10-1,000) ppm.[48]
The gas-oil fraction, which is the last distillate fraction of crude oil that is recovered, is the
feed to catalytic crackers. Metals such as iron,
copper, nickel, and vanadium contained in this

Table 7.3
Metal Content in Crude Oils

	Concentration Range (ppm)	Mean[a] or Median[b] Concentration (ppm)	Tons/Year (calculated from previous column)
Antimony	0.030-0.107	0.055 ± 0.003[a]	31
Arsenic	0.046-1.11	0.263 ± 0.007[a]	148
Barium	Small amounts in Texas crudes[43]	--	--
Cadmium	--	0.03[c]	17
Chromium	0.0016-0.017	0.008 ± 0.003[a]	4.8
Cobalt	0.032-12.751	1.71 ± 0.11[a]	963
Copper	0.13-6.33	1.32 ± 0.01[a]	743
Iron	3.365-120.84	40.67 ± 2.48[a]	22,900
Lead	0.17-0.31[44]	0.24[b]	135
Manganese	0.63-2.54	1.17 ± 0.04[a]	659
Mercury	0.023-30	3.24 ± 0.01[b]	1,820
Molybdenum	0.008-0.053[45]	0.031[b]	17
Nickel	49.1-344.5	165.8 ± 7.1[a]	93,400
Selenium	0.026-1.396	0.53 ± 0.044[a]	300
Silver	Traces in many crudes[43]	--	--
Tin	Identified in Mexican crudes		
Vanadium	4.0-298.5	88.55 ± 0.42[a]	49,900[d]
Zinc	3.571-85.80	29.80 ± 1.27[a]	16,800
		Total	187,837.8

[a]from references 41 and 42.
[b]from references in concentration range column.
[c]estimated from zinc content.
[d]W. E. Davis estimated 17,000 tons/year emissions from burning residual fuel oil.[46]

Table 7.4

Percentage Yields of Refined Petroleum Products From Crude Oil in the U.S. (1970)[1]

Product	%
Gasoline	45.3
Jet fuel	7.5
Ethane (including ethylene)	0.2
Liquefied gases	3.0
Kerosine	2.3
Distillate fuel oil	22.4
Residual fuel oil	6.4
Petrochemical feedstocks	2.5
Special naphthas	0.8
Lubricants	1.6
Wax	0.2
Coke	2.7
Asphalt	3.6
Road oil	0.3
Still gas	4.1
Miscellaneous	0.3
Shortage	-3.2

Table 7.5

Trace Metals in Distillate Fuels[47]

Fuel	V	Pb	Cu
Kerosines	0.0-3	0-3.0	0-4
Diesel fuels	<0.01-0.5	<0.01-5.0	<0.01-1.0
Burner fuels	0-0.07	0-3.5	0.002-0.42
Aviation turbine fuels	<0.01-0.05	<0.05-2.0	<0.01-0.10
Gas turbine fuels	<0.01-0.10	<0.02-2.0	<0.01-0.10

fraction poison the cracking catalysts.[49]   A nickel
equivalent* of <0.3 ppm is desired to insure good
catalyst life and performance.

It appears that only a fraction of the metals
found in the distillates comes from the crude oil.
A substantial portion is likely introduced during
storage and handling.

Although the metals that could be added to the
environment from distillates would appear to be substantial, the annual contribution to natural background levels would be small. A simple calculation,
based on (a) the estimates of metal contents of
crude oils consumed in the U.S. given in Table 7.3,
(b) the assumption that metals accumulate in only the
top inch of soil, and (c) the assumption that the
metal is uniformly introduced over the geographical
area of the U.S., shows an increase in the soil metal
content of 0.4 ppm.

The limited information available indicates,
however, that probably 90% of the metals in crude oil
are retained in the distillate bottoms, *i.e.*, residual
fuel oil and asphalt. These materials are a much
more concentrated source, and are distributed and
used in a much smaller geographical area.

Approximately 38% of the residual fuel oil produced annually is burned in power plants.[50] The
remainder is used for a variety of other industrial
and private uses. In industrial facilities a substantial fraction of the metals found in residual fuel
oil is collected in the bottom ash and in fly ash produced during its combustion. Some of the more volatile metals and metal oxides are lost to the environmen
as vapors together with some of the very fine particu-
lates, which are difficult to collect. Again, however,
no data presently available describe the ultimate fate
of metals in residual fuel oils.

The bottom ash is used largely as aggregate for
road construction, concrete production, and other
similar purposes. Fly ash is of less commercial value
and often is discarded directly to the environment.
The metals found in ash occur predominantly as the

---

*"Nickel equivalent" is a multiplier factor that describes
the relative effect of iron, copper, nickel, and vanadium
on cracking catalysts. The multipliers for the metals are
nickel 1.0, copper 1.0, vanadium 0.2, and iron 0.1. Thus,
a nickel equivalent of 0.3 ppm indicates that the sum of the
metal concentrations times their multiplier factor should
not exceed 0.3.

oxides. Since these compounds are generally very
susceptible to solubilization and leaching, they
have a ready mode of transport to surface waters.
Residual fuel oil, which can be characterized
as a "liquid coal," has a much lower sulfur content
than coal and, consequently, is in increasing demand
as an energy source. Catalytic processes are being
developed to reduce further the sulfur content of
this material. These processes reportedly remove a
large fraction (perhaps as much as 75%) of the
residual metals.

Nickel and vanadium removed during catalytic
processing will constitute a high-grade ore and could
supply a substantial amount of the domestic demand
for these metals. The U.S. consumption of nickel
and vanadium in 1970 was estimated at 155,719 and
5,134 short tons, respectively.[1] Crude oil, based
on our estimate, contains up to 60% of the annual
U.S. demand for nickel and nine times the annual
demand for vanadium.

The largest percentage of metals in crude oil
apparently collects in the asphalt fraction. This
material is used primarily in road construction and
for roofing. The available data on the metal contents
of asphalt are shown in Table 7.6. The minimum and
maximum amounts of metals per inch of asphalt per
mile of two-lane highways are also shown in Table
7.6.

No data have been found on the amount of metal
that leaches out of asphalt to enter the environment.
Urban runoff, even from highly domestic areas, contains unexpected large quantities of metals.[52]
Studies are needed to determine whether abrading of
asphalt streets and roofs is a significant source
of these metals.

USED PETROLEUM PRODUCTS

Although crude petroleum contains substantial
quantities of metals, the largest sources of metals
entering the environment from petroleum products are
tetraethyl lead added to gasoline, metal compounds
added to lubricants, and wear metals that accumulate
in used lubricants. The fate of lead in gasoline
has been studied extensively and will not be discussed
here.

Lubricants produced in a refinery are high-grade
products that retain little or no residual metal from
the crude oil. The demand for improved performance
characteristics for lubricants has been met by

Table 7.6

Metal Content of Asphalt and Asphalt Concrete

Metal	Concentration in Asphalt[45] (ppm)	Metal Content (lb/in) of Thickness/Mile of Asphalt Concrete of Highway[a]
Antimony	0.04-11.0	0.004-1.0
Chromium	0.5-3.6	0.04-0.3
Cobalt	0.05-0.14	0.004-0.01
Copper	0.5-11.1	0.04-1.0
Manganese	0.3-0.4	0.03-0.04
Molybdenum	2.8-10.1	0.2-0.9
Nickel	86-104	7.7-9.3
Vanadium	14-200	1.2-17.8
Zinc	0.4-1.8	0.4-0.2

[a]Based on 24-ft. wide asphalt pavement containing 6% asphalt (94% aggregate) with a pavement density of 140 lb/ft.^{3}

developing additives that prolong the useful life of lubricants. These additives encompass a wide range of chemical compounds, including metal salts added to reduce wear, resist oxidation, inhibit corrosion, and function as detergents, bactericides, antistatic agents, etc.

We were not able to develop complete data on how much of each compound is added to lubricants. The additives are sold as a package by a few firms and are designed for an individual consumer. Table 7.7, however, shows some of the types of metal-containing additives now in use.

The other major sources of metals in waste lubricants are those resulting from wear or use. The disposal of waste lube oils has been recognized as a potential pollution problem for many years. However, not until 1966 was the first attempt made to bring the problem into focus: service stations throughout the United States were surveyed to determine how they disposed of their waste automotive oils. The results of that survey led to a better awareness of the pollution potential of used lubricants and to the eventual formation of an American Petroleum Institute Task Force on Used Oil Disposal.

Table 7.7
Lubricant Additives*

Metal	Representative Compounds	Purpose
Antimony	Sb dialkyl dithiocarbamates	Antiwear, extreme pressure, and antioxidant additives in conventional and low-ash-type automotive crankcase oils, industrial and automotive gear oils, greases (amounts $\leq$ 1-3%)
Barium	Ba diorgano dithiophosphates Ba petroleum sulfonates Ba phenolates Ba phosphonates or thiophosphonates	Corrosion inhibitors, detergents, rust inhibitor, automatic transmission fluids, greases
Boron	Borax, boric acid esters	Antiwear agents, antioxidant, deodorant cutting oils, greases, brake fluid
Cadmium	Cd dithiophosphates	Steam turbine oils
Chromium	Cr salts	Grease additive
Lead	Pb naphthenate	Extreme pressure additive, greases, gear oils
Mercury	Organic mercury compounds	Bactericide, e.g., cutting oil emulsions
Molybdenum	MoS2-Mo dibutyl dithiocarbamate and phosphate	Greases, extreme pressure additives
Nickel	Cyclopentadienylnickel complexes	Antiwear agents, minimize carbon deposits, improve lubrication and combustion
Selenium	Selenides	Oxidation and bearing corrosion inhibitors
Tin	Organotin compounds	Antiscuffing additive, metal deactivators
Zinc	Zn diorgano dithiophosphates Zn dithiocarbamates Zn phenolates	Antioxidant, corrosion inhibitors, antiwear additives, detergent, extreme pressure additives, in crankcase oils, hypoid gear lubricants, greases, aircraft piston-engine oils, turbine oils, automatic transmission fluids, railroad diesel engine oils, differential and wet brake lubricants

*from References 51-54

Each year some 2.5 billion gallons of lubricants are sold in the United States.[55] Table 7.8 gives a breakdown of lubricant uses. Eventually, large quantities of these lubricants are disposed of as wastes, although how much, exactly, is hard to determine. One American Petroleum Institute estimate is that about 450 million gallons of waste automotive oils are disposed of annually in the United States; other estimates range as high as 750 million gallons per year.[55] The disposal of so much material, whatever the exact figure, is obviously important because of its pollution potential.

Table 7.8

U.S. Consumption of Lubricating and Industrial
Oils and Greases: 1971*

	Greases (lb)	Oils (gal)
Automotive		1,084,819,000
	456,104,000	
Aviation		15,950,000
Industrial lubricating		734,792,000
	503,270,000	
Other industrial oils		390,872,000
Total (excludes exports)	959,374,000	2,226,433,000

*from reference 56

Only one comprehensive study of waste lubricant disposal has apparently been done.[57] This study, by Arthur D. Little, Inc., showed the ultimate fate of used automobile oil in the State of Massachusetts (Table 7.9).

Reprocessed oil includes used oil re-refined and oil partially cleaned and used for fuel oil. The data on reprocessed oil, if accurate for Massachusetts, are not representative of national practices. It is estimated that the 1972 U.S. re-refining capacity was only 100 million gallons,[55] or less than 4% of the U.S. consumption.

Table 7.9

Disposal of Waste Automobile Oil*

Means of Disposal	*Percentage*
Reprocessed for fuel oil	36.6
Road oil use	11.4
Taken out of state	8.1
Dumped on ground at source	22.8
Dumped into sewer at source	0.8
Farm use	2.4
Fate unknown	17.9

*from Reference 57

Disposal of used lubricants on roads or on parking lots continues to be a wide practice, even though the Environmental Protection Agency has recently demonstrated that waste lubricants are only marginally effective as a means of dust control on dirt or gravel roads. Of more environmental importance, however, is the fact that this material is high in emulsifiers and resistant to oxidation under atmospheric conditions.[12] Thus, during storms it is often washed off the road into the ground.

If waste-lubricant disposal is subdivided into two general categories—dumping and reprocessing—it appears that as much as three-fourths of the nation's used automotive oils may be disposed of by dumping.

Table 7.10 shows the typical metal content of used automobile engine oil. On the basis of an estimate of 1.1 billion gallons of lubricant used for this purpose,[61] we have calculated the quantity of metals entering the environment from waste automotive oil. The 29,000 tons of lead calculated to be found in waste lubricants amounts to only 10% of the lead added to gasoline each year.[39]

About the same amount of lubricant is used by industry for such purposes as machining oil (grinding, milling, cutting, and turning); hydraulic lifts; and lubricants in cold strip rolling. Many of these industrial lubricants also serve as coolants. These materials run the gamut from high water-to-oil emulsions to esoteric water-soluble oils. These oils are commonly used several times and have life cycles varying from a few days to several weeks.

Table 7.10

Metals in Used Engine Oils*

	$Lb/10^6$ Gal			
Metal	In Combustion Products of Crankcase Drainings[a]	Used Reciprocating Engine Oils[b]	Used Jet Engine Oils[c]	Total Amounts In Automotive Lubricants (short tons)[d]
Barium	3,000	--	--	1,650
Boron	130	--	--	72
Chromium	180	53	22	99
Copper	120	103	67	66
Iron	2,200	245	210	1,210
Lead	53,000	24,000[e]	142	29,150
Nickel	50	10.7	8.6	28
Silver	--	12.3	10.5	--
Tin	71	15	2-15	39
Zinc	3,400	--	--	1,870

*from References 59 and 60.
[a] Average of nine cities.[23] From data compiled and published by Walter C. McCrone Associates, Inc., Chicago, Illinois.
[b] Average, three to five samples (Air Force data).[24]
[c] Average, 13 to 17 samples (Air Force data).[24]
[d] Based on Column 1.
[e] Mean of range medians.

Insufficient data on metal content of these waste oils are available since modes of usage are so varied. It is apparent, however, that the metal content of these materials can be very high.

Increasingly industries are often reprocessing and recycling their own oily waste in-house. Where recycling is practiced, the metal contaminants are separated, probably lagooned, and land-disposed or returned to surface waters.

To summarize, there simply are not enough reliable data on the total amount of metal lost to the environment each year from processing, use, and disposal of petroleum products. We know that it is large, but one must keep the problem of environmental contamination in perspective.

We are still in the early stage of understanding the relationship of most metals to human health. Until we have greater knowledge, we must be cautious about contaminating the environment with any metals until we gain a better understanding of their physiological role, their mode of transport in the environment, and their pathways to the food chain.

REFERENCES

1. Bureau of Mines Staff, *Minerals Yearbook, Vol. I, Metals, Minerals, and Fuels 1970*, Bureau of Mines, U.S. Department of the Interior (Washington, D.C.: U.S. Government Printing Office, 1972), pp. 783, 822, 888, and 1163.
2. Underwood, E. J., *Trace Elements in Human and Animal Nutrition* (New York: Academic Press, 1971).
3. Shroeder, H. A., *Pollution, Profits and Progress* (Brattleboro, Vermont: The Stephen Green Press, 1971).
4. Lee, D. H. K., ed., *Metallic Contaminants and Human Health* (New York: Academic Press, 1972).
5. Stokinger, H. E., *Am. Ind. Hyg. Assoc. J.*, 30 (3), 195 (1969).
6. "Huge Check on Chemicals as Carcinogens," *Chem. Eng. News*, 50 (2), 6 (1972).
7. O'Dell, B. L. in *Proceedings, University of Missouri's 1st Annual Conference on Trace Substances in Environmental Health*, Columbia, Missouri, July 10-11, 1967 (1968), pp. 134-140.
8. Snively, W. D., Jr., and B. Becker, *Ann. Allergy*, 26, 233 (1968).
9. Stecher, P., M. J. Finkel, O. H. Siegmund, and B. M. Szafranski, eds. *The Merck Index of Chemicals and Drugs*, 7th ed. (Rahway, New Jersey: Merck and Company, Inc., 1960).
10. Schroeder, H. A., *Arch. Environ. Health*, 21, 798 (1970).
11. Sax, N. I. *Dangerous Properties of Industrial Materials*, 2nd ed. (New York: Reinhold Publishing Corporation, 1963).
12. Browning, E., *Toxicity of Industrial Metals* (London: Butterworths & Company, 1961).
13. Hwang, J. Y., *Anal. Chem.*, 44 (14), 20A (1972).
14. Louria, D. B., M. M. Joselow, and A. A. Browder, *Ann. Internal Med.*, 76, 307 (1972).
15. Gafafer, W. M., ed., *Occupational Diseases, A Guide to Their Recognition*, U.S. Department of Health, Education, and Welfare, Public Health Service (Washington, D.C.: U.S. Government Printing Office, 1964).
16. Arthur D. Little, Inc., *Water Quality Data Book, Vol. 2, Inorganic Chemical Pollution of Freshwater*, Water Pollution Control Research Series, Environmental Protection Agency (Washington, D.C.: U.S. Government Printing Office, 1971).

17. Lisk, D. J., in *Advan. Agron.*, Vol. 24 (New York: Academic Press, 1972), pp. 267-325.
18. Scanlon, J., *Clin. Pediatrics* (Philadelphia), *11*, 135 (1972).
19. Albright, L. J., J. W. Wentworth, and E. M. Wilson, *Water Res.*, *6*, 1589 (1972).
20. Bowen, H. J. M., *Trace Elements in Biochemistry*, (London: Academic Press, 1966).
21. Miner, S., "Air Pollution Aspects of Barium and Its Compounds," National Technical Information Service Document PB 188 083, 1969.
22. Dooley, A. E., *Arch. Environ. Health*, *6*, 30 (1963).
23. Taylor, F. B., *J. American Water Works Assoc.*, *63* (11), 728 (1971).
24. Sauchelli, V., *Trace Elements in Agriculture* (New York: Van Nostrand Reinhold, 1969).
25. Murphy, L. S., and L. M. Walsh, in *Micronutrients in Agriculture*, J. J. Mortvedt, P. M. Giordano, and W. L. Lindsay, eds. (Madison, Wisconsin: Soil Science Society of America, Inc., 1972), p. 347.
26. "Trace Metals: Unknown, Unseen Pollution Threat," *Chem. Eng. News*, *49* (29), 29 (1971).
27. "Earthworms High in Metals," *Chem. Eng. News*, *50* (41), 60 (1972).
28. John, M. K., C. J. VanLaerhoven, and H. H. Chuah, *Environ. Sci. Technol.*, *6* (12), 1005 (1972).
29. Page, A. L., F. T. Bingham, and C. Nelson, *J. Environ. Quality*, *1* (3), 288 (1972).
30. Sartor, D., and G. B. Boyd, "Water Pollution Aspects of Street Surface Contaminants," Environmental Protection Technology Series, EPA-R2-72-081 (November 1972).
31. Lanford, C. E., *Oil Gas J.*, *67* (13), 82 (1969).
32. Mukherji, S., and B. D. Gupta, *Physiol. Plant.*, *27*, 126 (1972).
33. Bunch, L. D., and C. S. Reusch, *J. Kansas Med. Soc.*, *69* (7), 339 (1968).
34. Pletnikova, I. P., *Gig. Sanit.*, *35* (2), 14 (1970); *Hygiene and Sanitation*, *35* (1-3), 176 (1970).
35. Lemley, R. E., and M. P. Merryman, *J. Lancet*, *59* (11 New Series), 435 (1941).
36. Hoffman, I. (Symposium Chairman), "International Symposium on Identification and Measurement of Environmental Pollutants," National Research Council, Ottawa, Canada (1971).
37. Pendergrass, E. P., and A. W. Pryde, *J. Ind. Hyg. Toxicol.*, *30* (2), 119 (1948).
38. Lagerwerff, J. V., U.S. Department of Agriculture, personal communication, 1973.
39. Hall, S. K., *Environ. Sci. Technol.*, *6* (1), 31 (1972).
40. "And Now, Cadmium," *Time*, *97* (10), 35 (1971).
41. Shah, K. R., R. H. Filby, and W. A. Haller, *J. Radioanal. Chem.*, *6*, 185 (1970).

42. Shah, K. R., R. H. Filby, and W. A. Haller, *J. Radioanal. Chem.*, *6*, 413 (1970).

43. Thomas, W. H., *Inorganic Constituents of Petroleum, Science and Petroleum*, Vol. 2, A. E. Dunstan, ed. (New York: Oxford University Press, 1938) p. 1053.

44. Bratzel, M. P., Jr., and C. L. Chakrabarti, *Anal. Chim. Acta*, *61* (1), 25 (1972).

45. Colombo, U. P., G. Sironi, G. B. Fasolo, and R. Malvano, *Anal. Chem.*, *36* (4), 802 (1964).

46. Davis, W. E., and Associates, "National Inventory of Sources and Emissions. Arsenic, Beryllium, Manganese, Mercury and Vanadium. 1968. Vanadium. Section V," Report for Environmental Protection Agency, Office of Air Programs (1971).

47. Ward, C. C., "Survey of Trace Metals in Distillate Fuels," presented at the Symposium of Gas Turbine Fuel Requirements, Handling and Quality Control, June 28-29, 1972, Los Angeles, California.

48. Lee, R. E., Jr., and D. J. von Lehmden, *J. Air Pollution Control Assoc.*, *23* (10), 853 (1973).

49. John Winters, American Oil Company, personal communication (August 1973).

50. Davis, W. E., and Associates, "National Inventory of Sources and Emissions, Arsenic, Beryllium, Manganese, Mercury and Vanadium 1969. Selenium. Section IV," Report for Environmental Protection Agency, Office of Air Programs, 1971.

51. Smalheer, C. V., and R. Kennedy Smith, *Lubricant Additives* (Cleveland, Ohio: Lezius-Hiles, 1967).

52. Farmer, H. H., B. W. Malone, and H. F. Tompkins, *Lubrication Eng.*, *23* (2), 57 (1967).

53. Sullivan, R. J., "Air Pollution Aspects of Nickel and Its Compounds," National Technical Information Service, PB 188 070, 1969.

54. Asseff, Peter A., *Lubrication Eng.*, *23* (3), 107 (1967).

55. American Petroleum Institute, "Waste Oil Roundup, No. 1," Committee on Disposal of Waste Products, Division of Marketing, API, Washington, D.C. (1972).

56. Bureau of the Census, "Sales of Lubricating and Industrial Oils and Greases, 1971," in "Current Industrial Reports," U.S. Department of Commerce (October 1972).

57. Arthur D. Little, Inc., "A Study of Waste Oil Disposal Practices in Massachusetts," A Report to the Commonwealth of Massachusetts, Division of Water Pollution Control, January 1969.

58. Pendleton, Harold E., Manager, Environmental Services Committee on Public Affairs, American Petroleum Institute, Washington, D.C., personal communication (December 7, 1972).

59. Booth, G. T., Jr., "The Oil Company's Partner in Proper Service Station Waste Oil Disposal--the Collector and Re-refiner," Presented at the National Fuels and Lubricants Meeting, September 14-15, 1972, New York, N.Y.

60.   Reeves, R. D., C. J. Molnar, M. T. Glenn, J. R. Ahlstrom, and J. D. Winefordner, *Anal. Chem.*, *44* (13), 2205 (1972).
61.   "Waste Lube Oils Pose Disposal Dilemma," *Environ. Sci. Technol.*, *61* (1), 25 (1972).

CHAPTER 8

TRACE ANALYSIS OF IRON, NICKEL, COPPER AND
VANADIUM IN PETROLEUM PRODUCTS

A. J. Smith, J. O. Rice, W. C. Shaner, Jr. and C. C. Cerato
Sun Oil Company
P.O. Box 1135
Marcus Hook, Penn.

The significant effect of traces of iron, nickel, copper and vanadium in petroleum feedstocks on processing economics has been responsible for the continuing attention focused on improvements in trace metal analytical procedures. This interest has intensified in our company because of the wide variety of crude oils now processed coupled with the need to meet more stringent contamination factors. These needs have increased demands on our laboratories for routine analyses at levels of less than 1 ppm. As a consequence our group was asked to evaluate requirements for trace metal analysis and to determine to what extent methods presently described in the literature could be adapted to our needs.

Since a standard Philips X-ray spectrograph and a Perkin-Elmer 403 Atomic Absorption spectrophotometer were available to us, we evaluated procedures using these techniques after incorporating a number of improvements suggested in the literature. It should be noted at this point, however, that these instrumental techniques did not always yield the quality of analytical results that might be expected. Our purpose in presenting somewhat imperfect data is to give a realistic view of what might be expected when the samples do not conform to the techniques at hand.

Attention was first given to possible improvement of our routine XRF methods. Of the number of methods in the literature dealing with XRF analysis the work of three groups shown in Table 8.1 was of particular interest. The first paper by the group from Gulf[1]

149

Table 8.1

Analysis of Three XRF Analytical Methods

Author	Metals	Sample Preparation	XRF Measurement
J. E. Shott, et al.	Ni, V	Ash 10-g sample with 3 g benzene sulfonic acid and Co internal standard.	Measure NiKα/CoKβ and VKα/CoKβ in ash on filter disk.
W. A. Rowe, et al.	Ni, V Fe, Cu	Ash 50-g sample with xylene sulfonic acid and cobalt internal standard.	Paste ash to mylar film and measure Kα first order line of Co, Cu, Fe, Ni and V.
J. G. Bergmann, et al.	Ni, V Fe	Ash 10-g sample with sulfur. Dissolve residue and absorb metals on ion exchange disk.	Count each element for 100 sec using Kα line.

served   as an excellent starting point for the analysis of trace metals by XRF. These workers reported that losses of volatile porphyrins could be obviated by ashing in the presence of benzene sulfonic acid, and they suggested that addition of cobalt as an internal standard for ashing would enhance reproducibility.

The next paper by Rowe and Yates[2] described an extension of the above method that included Cu and Fe as well as Ni and V with improved precision for the analysis of nickel.

In a more recent account Bergmann[3] and co-workers adapted the use of sulfur as a reagent to reduce porphyrin loss during combustion, thus obviating iron contamination from benzene sulfonic acid. Also described was a procedure for the concentration of metals on a suitable disk of ion exchange paper. This technique offered a distinct advantage in that the thin layer gives a high yield of secondary emission Also the background scatter is reduced because the sample consists of metals only on the disk. Using XRF further refined with curved crystal optics they reported good accuracy and precision for nickel and vanadium in synthetic unknowns ranging from 0.1 to 1 ppm.

In this paper we will describe our effort to evolve a method that would incorporate such features as:

1. applicability to standard instrumentation
2. applicability to Fe, Ni, Cu and V
3. use of an internal standard
4. use of a 10-g sample

Initially for routine work we adopted the Bergmann method. Table 8.2 shows the results of our analysis of a synthetic unknown prepared by diluting Boscan crude 650 bottoms in 94 Golden oil. Considering that the calculated value was based on the average of analyses from a number of laboratories that used a variety of methods for analysis we found the agreement with the expected value reasonable. However, we thought that our precision might possibly be improved if mechanical loss due to ashing could be compensated by the use of an internal standard.

Table 8.2

XRF Analysis of Boscan Crude 650 Bottoms Diluted with Transformer Oil

Metal	Expected	Found	σ	DF
Fe	1.76	1.52	0.35	4
Fe	3.44	2.72	0.35	4
Ni	1.56	1.82	0.15	4
V	1.65	1.54	0.33	5
Cu	1.50	1.76	0.33	5
Cu	1.40	1.16	0.33	5
Ni	0.34	0.48	0.28	4
V	0.37	0.21	0.05	4
Cu	0.30	0.50	0.08	3

Table 8.3 demonstrates the use of an internal standard when the X-ray intensities of Fe, Ni, Cu and V are measured relative to cobalt on ion exchange disks. These samples were derived from solutions of pure salts prepared for calibration. The half disk measurements give an indication that the ratios do not change significantly when only one-half the amount of metal is available for measurement. If these measurements were derived from 10-g samples, agreement would be quite adequate.

Table 8.3

Effect of Internal Standard on Ion Exchange Separation

	μg* Added	Fe Ratio	Fe μg	Ni Ratio	Ni μg	Cu Ratio	Cu μg	V Ratio	V μg
Whole disk	4	0.145	5.2	0.246	3.5	0.695	3.7	0.550	4.1
Half disk		0.154	7.3	0.305	4.9	0.902	6.8	0.743	7.2
Whole disk	12	0.232	12.5	0.569	11.0	1.196	11.3	1.041	12.1
Half disk		0.243	13.4	0.590	11.6	1.321	13.2	1.131	13.6
Whole disk	20	0.329	20.8	0.931	19.6	1.777	19.8	1.314	20.2
Half disk		0.307	17.0	0.817	17.0	1.864	21.1	1.609	21.8

*Added as solutions of pure salts.

Table 8.4 shows the results that were obtained by analyzing trace metals on ion exchange disks. Since these disks were prepared by treatment with solutions of metal salts of known concentration, the precision data give an estimate of the error associated entirely with the ion exchange and measurement steps of the method.

Table 8.4

Recovery of Metals on Ion Exchange Disk
Using Internal Standard

Level μg *Added**	*Ni* $\overline{X}$	σ	*Cu* $\overline{X}$	σ	*V* $\overline{X}$	σ	*Fe* $\overline{X}$	σ
4	3.94	0.069	4.39	0.71	4.28	0.094	3.90	0.52
8	8.29	0.25	8.35	0.86	8.29	0.22	7.70	0.75
12	12.40	0.25	12.82	0.82	12.26	0.24	11.84	1.20
16	16.32	0.35	16.43	0.58	16.08	0.37	15.14	2.16
20	20.01	0.53	20.63	0.73	17.88	0.26	18.97	2.14

*Added as solutions of pure salts.

X = mean values from five determinations expressed in micrograms.

Table 8.5 shows the results that were obtained by analyzing a synthetic unknown prepared by diluting a Conostan D-20 standard in a mixture of 94 and 96 Golden oil. We chose this system because we felt that we could evaluate the accuracy of our data with greater reliability. As it turned out, we determined that the method becomes insensitive around the 0.1 ppm level. We also noted that the vanadium results were much lower than expected. Since we expected that the vanadium might be volatilized during the ashing step, we prepared a synthetic unknown using our analyzed sample of Boscan 650 bottoms. Our results showed better recovery of vanadium, which to us indicated that Conostan vanadium might be somewhat volatile.

Our results also serve as a reminder that although the use of an internal standard can offer some protection against mechanical losses, it cannot compensate for losses of metal due to the volatility of

the compound itself. Our results give further support to the axiom that the principal sources of
error are always associated with combustion.

Table 8.5

XRF Analysis of a Synthetic Unknown* Using an Ion
Exchange Disk with an Internal Standard

	1.00 ppm Level	σ	0.5 ppm Level	σ
Ni	0.77	0.24	0.45	0.12
Cu	1.22	0.45	0.66	0.26
Fe	0.61	0.21	0.34	0.10
V	0.36	0.14	0.26	0.16

*Conostan D-20 Standard in a mixture of 94 and 96 Golden oil.

ATOMIC ABSORPTION

    Because atomic absorption enables direct analysis
without the need of a concentration step it held prom-
ise for metals ranging from 0.1 to 1 ppm. Table 8.6
shows the absorbances obtained for nickel in a synthetic unknown prepared from Boscan crude by the method
of Slavin and Trent.[4] An air-acetylene flame afforded
convenient combustion, and absorption at the 2316
non-absorbing line was subtracted from the sample
absorbance observations to correct for background absorption. The observations were made at the 100 average mode using the Nixie tube digital readout.
Figure 8.1 shows how these data were used to calculate
the nickel content of the synthetic unknown by the
standard addition procedure. Although these data gave
adequate results we found that the experiment was difficult to repeat since the background absorbance has
a tendency to increase with the amount of nickel added.
    Table 8.7 shows the results of our analysis of
Fe, Ni, Cu and V in MIBK. These solutions were prepared by diluting Conostan standards in MIBK only.
The precision and the accuracy of these analyses indicate the optimum since no correction for background

was necessary.   The standard addition calibration
curves were all quite linear.   The most irregularity
occurred for vanadium, where nitrous oxide-acetylene
combustion was used.

Table 8.6

*Atomic Absorption Analysis of Trace
Nickel by Standard Addition*[a]

μg Ni[b] *Added per ml*	*Absorbance at 2320A*	*Absorbance at 2316A*
0	0.028	0.014
0.05	0.031	0.014
0.10	0.035	0.015
0.20	0.041	0.013
0.30	0.050	0.016

a   from Reference 4.
b   Boscan 650°F fraction in 92 Golden oil.

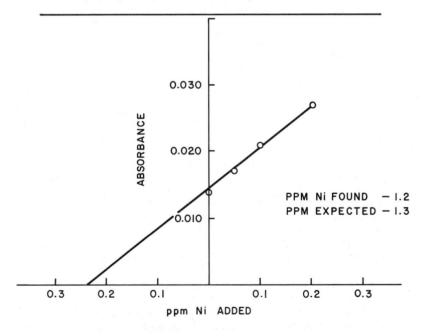

*Figure 8.1.   Analysis of trace nickel in Boscan crude,
synthetic unknown.*

Table 8.7

AA Analysis of Fe, Ni, Cu and V in
MIBK by Standard Addition

ppm Expected*	Fe	Ni	Cu	V
0.25	0.20,0.18	0.25,0.20	0.23,0.20	0.18,0.25
0.50	0.38,0.40	0.40,0.25	0.45,0.45	0.25,0.45
0.75	0.70,0.70	0.55,0.50	0.70,0.70	0.65,0.60
1.00	1.00,0.90	0.75,1.00	0.95,0.93	1.00,1.00

*Conostan   D-20 Standard diluted in MIBK.

Table 8.8 shows the background absorption of
three hydrocarbons considered as matrixes for the
preparation of synthetic unknowns. This study came
about as a result of an attempt to analyze Fe, Ni,
Cu and V in polyisobutylene from 0.1 to 1.0 ppm.
All the results were low except vanadium. Samples
containing only the matrix and solvent showed a
negative absorbance at the Fe, Ni, and Cu wavelengths.
The vanadium setting (nitrous oxide flame) showed a
zero absorbance.
       Table 8.9 shows the analysis of the synthetic
unknowns prepared in polyisobutylene. All four metals
were analyzed using a nitrous oxide-acetylene flame.
All the standard addition curves were linear; since
combustion of the petroleum matrixes with nitrous
oxide showed no effect we felt justified in not including a background correction at the non-absorbing
line.
       Table 8.10 shows the results obtained using the
nitrous oxide combustion techniques. For the Boscan
sample there was a pronounced difference in the
quality of data between the iron and copper analyses.
The linearity of the standard addition curves for
copper is reflected by the precision and the agreement with the expected value. The absorbance value
obtained for iron, on the other hand, yielded erratic
data points when plotted for the standard addition
calculation.
       We are not able to explain the value of 0.05 ppm
for nickel except to say that the data points of the
calibration curve from which the value was derived
were much more irregular than was the case for the
0.40 and 0.48 values.

Table 8.8

Atomic Absorption Background Effects of Petroleum Compounds

	Fe Wavelength = 2483A		NI Wavelength = 2320A		Cu Wavelength = 3247A		V Wavelength = 3184A	
	MIBK	Xylene	MIBK	Xylene	MLBK	Xylene	MIBK	Xylene
Propylene tetramer	-0.005	-0.052	-0.011	-0.048	-0.001	0.018	0.000	0.000
Polyisobutylene	-0.003	-0.036	-0.010	-0.036	0.000	-0.012	0.000	0.000
SunVis 7	0.000	-0.045	-0.008	-0.039	0.003	-0.010	0.000	0.000

Table 8.9

Atomic Absorption Analysis of Fe, Ni, Cu and V in
Polyisobutylene by Nitrous Oxide Combustion

Expected Value*	Fe	Ni	Cu	V
0.25	0.10,0.30	0.05,0.30	0.10,0.20	0.30,0.18
0.50	0.38,0.60	0.45,0.65	0.48,0.45	0.40,0.43
0.75	0.70,0.68	0.50,0.65	0.70,0.75	0.60,0.65
1.00	0.93,0.90	0.95,1.00	0.88,1.00	0.75,0.88

*Trace metals added as Conostan Standard.

Table 8.10

Atomic Absorption Analysis of Fe, Ni, Cu and V in
Petroleum Products Using
Nitrous Oxide Combustion

	Boscan #2		650 BTMS*	
	ppm Fe	ppm Cu	ppm Ni	ppmV
1	0.10	0.45	0.40	0.25
2	0.00	0.55	0.48	0.25
3	0.05	0.50	0.05	0.30
X	0.05	0.50	0.31	0.28
Exp	0.26	0.46	0.50	0.90

*Neill S/Odom Line Crude

The precision of the vanadium results was ac-
ceptable. The standard addition curves were linear,
and since the expected value is based on a single
determination its accuracy is suspect.

CONCLUSION

An attempt has been made to give a realistic account of our recent experience with AA and XRF for the analysis of Fe, Ni, Cu and V at the sub ppm level in typical petroleum products. While other workers have reported more precise results determined under certain restricted conditions, our results indicate that precise analysis at low levels is still not always possible. For XRF the concentration step still is a predominant source of error, while for atomic absorption matrix interference continues to impede sensitive metal detection. We call attention to these shortcomings first to encourage other workers to bring about improvement and second to alert those who are recipients of such analyses to be aware of the sources of error associated with such data.

REFERENCES

1. Shott, J. E., Jr., T. J. Garland and R. O. Clark. *Anal. Chem. 33,* 507 (1961).
2. Rowe, W. A. and K. P. Yates. *Anal. Chem. 35,* 368 (1963).
3. Bergmann, J. G., *et al. Anal. Chem. 39,* 1258 (1967).
4. Trent, D. and W. Slavin. *Atomic Absorption Newsletter, 3,* 131 (1964).

CHAPTER 9

THE OCCURRENCE OF MOLYBDENUM IN PETROLEUM

William K. T. Gleim, John G. Gatsis and Chester J. Perry
Universal Oil Products Company
Des Plaines, Illinois

Molybdenum occurs in petroleum, and has been found in some crudes, atmospheric tower bottoms and vacuum tower bottoms by emission analysis. We used this method in the beginning, but later switched to atomic absorption, which is more reliable. The smallest amount of molybdenum that can be determined by this method is 0.4 ppm. Therefore, it can be found more readily in vacuum or atmospheric tower bottoms of crudes with a low content of molybdenum. The analytical methods are described at the end of this chapter.

The data presented in Table 9.1 are the averages of five determinations. We have found molybdenum in four crudes--Athabasca Tar Sand Oil, Cold Lake, Lloydminster and Boscan. The first three are probably related geologically, the amount of molybdenum found in them decreasing with the increase in depth at which these crudes are found. By far the highest concentration occurs in Athabasca Tar Sand Oil, where most of the molybdenum seems to occur organically bound, as evidenced by the recovery of 70% of the molybdenum in the organic soluble fraction (see Figure 9.2). We have not investigated the nature of the organic complexing compounds complexing the molybdenum. Its presence in marine deposit[1] has been documented recently.

The occurrence of molybdenum in petroleum is too low to be of any consequence to refining. Molybdenum accompanies iron in several enzyme systems, xanthine oxidase, aldehyde oxidase, nitrate reductase and nitrogenase. These enzyme systems catalyze electron transfer. The most probable source

161

Table 9.1

Molybdenum Determinations

	ppm Mo
Crudes	
Athabasca Tar Sand Oil	10
Cold Lake	7.3
Lloydminster	3.3
Boscan	4.6
Atmosphere Bottoms	
Lagunas	0.7
Minas	-
Souedieh	3.9
Kuwait	<0.4
Vacuum Bottoms	
Wyoming Pipeline	1
West Texas	0.5
Poza Rica	4
Murbane	<0.4
Gach Saran	<0.4

of these enzyme systems is the nitrogen-reducing bacteria, which were carried in the oil-bearing strata by surface water percolating through the surface oil. The deeper the oil layer, the more the water was filtered and freed of bacteria by absorption through the soil; and the fewer the bacteria carried into the oil-bearing formation, the lower the molybdenum content of the oil. That is the reason why the Athabasca Tar Sands, which are close to the surface, contain the highest amount of molybdenum. Whether the nitrogen-fixing bacteria contributed to the nitrogen content of the oils by nitrogenation has to be left to future investigation.

The prosthetic groups of these enzyme systems
contain flavin nucleotides, Figure 9.1. The flavin
nucleotides contain the isoalloxazine and the purine
ring system. Both of these ring systems are nitrogen heterocycles and are quite heat stable. Therefore
it should be possible to find both isoalloxazines
and purines among the nitrogenous compounds of crude
oils. It might be these compounds that complex the
molybdenum in oil. Of course even isolation of these
compounds could not prove if nitgrogen-fixing bacteria
are involved in the petroleum maturation process.

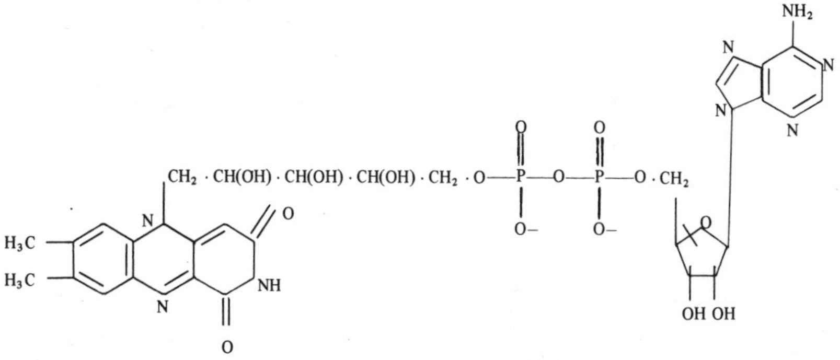

Figure 9.1. Flavin adenine dinucleotide (FAD).

DETERMINATION OF MOLYBDENUM BY ATOMIC
ABSORPTION SPECTROPHOTOMETRY (AAS)

Sample Preparation

To insure that all the molybdenum in the oil
whether present in an organically soluble form or in
varying forms as particulate matter would be recorded
by AAS, a sulfuric and nitric acid wet digestion process was selected for sample preparation. This process differs from the wet ashing procedure in that
the former is carried out in a liquid acid medium
from start to finish at a relatively low temperature
(280°C), whereas in the wet ashing procedure after
acid charring the oil, the char is subjected to a
600°C muffling step.

Separating Athabasca Tar Sand Oil

Three hundred grams of warm Athabasca Tar Sand
Oil were poured into 900 ml agitated n-heptane and
stirred 30 minutes. The mixture was warmed in a
steam bath to 70°C, stirred again and centrifuged
for 30 minutes at 3500 rpm. The insolubles were
restirred with n-heptane and centrifuged. The liquids
were combined and the solvents removed on a rotary
flash evaporator. The solids were dried, then re-
fluxed with 300 ml benzene for 1.5 hours, stirred for
30 minutes, and centrifuged. The solids were reslur-
ried three times with benzene as above, then dried in
a vacuum oven at 50°C at 20 torr. The benzene extracts
were combined and the solvent removed on rotary flash
evaporator (Figure 9.2).

Procedure

Two portions of oil, 55 g ± 0.1 g, were wet-
digested separately with sulfuric and nitric acids
in a new in-house glass laboratory digestion apparatus.
Prior to wet digestion, 1000 γ of a Conostan molybdenum sulfonate was added to one of the portions of
the oil. After digestion, these two portions of the
digested oil would now represent a modification of
the classical AAS method of standard additions with
one added desirable feature: the added molybdenum
standard would be subjected to the same harsh acidic
treatment as any of the molybdenum present in the
original oil. We refer to this procedure as the
"cooked" method of standard additions, and find it to
be the ultimate in matching atomic absorption response
between the sample and the sample plus the added standard. A blank acid digest solution was also carried
along.
The sulfuric acid in the finished digest solution
was fumed off to approximately 10 ml and, after water
dilution and heating, was quantitatively transferred
to a 100-ml flask. This produced a digested 55-g
sample in one flask and, in the other flask produced
a similar 55-g sample plus an added concentration of
10 γ/ml of molybdenum standard. These solutions
without any further treatment were aspirated into
AAS flame.

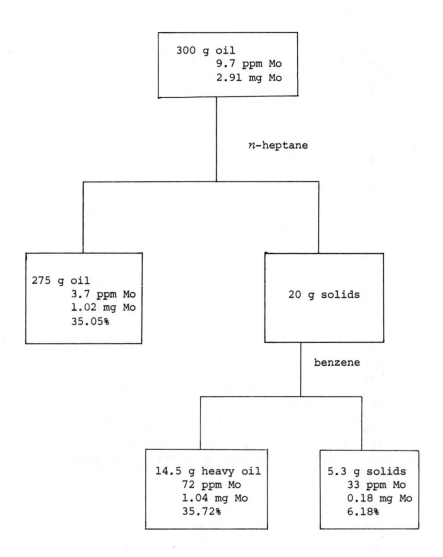

Figure 9.2. *Separation of Athabasca Tar Sand Oil.*

Instrumental Parameters

A Jarrell-Ash Model 82-532 MV spectrophotometer
equipped with a Perkin Elmer nebulization system and
a Leeds-Northrup Type W calibrated AZAR Recorder was
used for the molybdenum analysis. The 3133 Å primary
molybdenum absorption wavelength coupled with an
acetylene-nitrous oxide flame were the basic parameters
of the instrument. The recorder was set for a 5-fold
signal expansion along with the amplifier at 3/4
damped position.
    The blank acid digestion solution was checked
for response and found to produce a less than 0.2%
absorption; therefore deionized water was aspirated
as a zero reference. The 100-ml digest solutions
were then aspirated and their responses recorded.
The minimum signal accepted for calculation was 0.5%
absorption, or slightly more than twice the noise observed while aspirating the digest solutions. Ioniza-
tion supressant agents, such as sodium or potassium,
were not employed because we wanted to observe if
different crudes upon digestion recorded varying absorption responses for the molybdenum standard that
was added.

Results and Conclusions

    Almost all the digested crudes produced absorption
responses between 33 and 40% for the 10 ppm of molybdenum added. One exception was Souedieh Reduced Crude-
its response for the 10 ppm Mo added was 18% absorption
No two crudes produced identical responses. It was
apparent from some digested crudes that larger than
55-g samples could be digested, which would lower detection limits. The addition of sodium or potassium
at a 1000 γ/ml level also would increase response.
    The digested "cooked" method of standard additions
appears to be the proper method for AAS sample preparation especially if unfamiliar reduced crudes are to
be analyzed for trace amounts of molybdenum (<10 ppm)
for the first time.

REFERENCES

1. Bertine, K. K. and K. K. Turekian. *Geochim. Cosmochim.
    Acta 37,* 1415 (1973).

CHAPTER 10

VANADIUM AND ITS BONDING IN PETROLEUM

T. F. Yen
Department of Chemical Engineering
University of Southern California
Los Angeles, California 90007

INTRODUCTION

Vanadium is ubiquitous in nature. All types
of igneous rocks and soils contain an average of
100 ppm vanadium, although its concentration in
sea water is much lower--about 0.002 ppm. All
living organisms contain vanadium from 0.15 to 2
ppm; certain land plants such as *Amanita muscaria*
and marine animals such as *Pleurobranchas plumula*
contain vanadium in concentrations as high as 0.65%
on a dry weight basis.[1] Biochemically vanadium
serves as an essential pigment in blood and is involved in processes ranging from nitrogen fixation
to inhibition of cholesterol synthesis. Its importance is further supported by the fact that
vanadium is linked to RNA and DNA, the latter being
essential to the life process.

Vanadium is known to associate with the heavy
end fractions of crude oil in amounts ranging from
1 to 6,000 ppm. Since these vanadium sites are
extremely stable, they become the centers for cata-
lyzing oxidation reactions (transformation and diagenesis) in petroleum as well as adverse effects
in the refining processes such as catalytical cracking. Vanadium contributes to harmful physiological
effects, such as lung disease, arising from the industrial combustion of fuels and the resultant ejection of vanadium derivatives into the atmosphere.
Heterocyclic atoms at the vanadium sites are essential in keeping the homogeneity of petroleum oil. Many

167

physical properties, such as flow and viscosity, depend on the geometrical arrangement of the vanadium sites. There are a number of types of vanadium complexes in petroleum, the commonest being the porphyrins. Much work has been directed to detect their presence since they are readily separated by extraction. Other types of complexes, such as mixed ligands, pseudoaromatic pheophorbides and highly aromatic porphyrins, comprise 1-50% of the total vanadium.[2] These complexes are held tightly in the organic matrix and thus exert strong influence on the properties of petroleum samples. In this important aspect, few studies have been made (see Figure 10.1 for details).

Vanadium macrocycles could serve as geochemical tracers that link geo- and biosystems. Vanadyl porphyrins have been identified both in Precambrian shale[3] as well as in extraterrestrial meteorites. Although there are reports of the presence of either vanadium or porphyrin-like molecules in many types of recent and ancient sediments, the nature of vanadium bonding in all these samples is not known. Once this knowledge is acquired, an insight into the structure of these natural resources will facilitate their intelligent applications in energy conversion and environmental control.

METHOD

Electron spin resonance (ESR) is an important tool in the exploration of the microenvironment where there are large molecules in their natural state without chemical decomposition or physical deformation. Furthermore, solubilization of the substance is unnecessary, thus allowing spectra to be taken *in situ*. This technique appears to be most desirable for the proposed study. Previously this investigator has employed ESR techniques to study the nature of free radicals and vanadyl porphyrin complexes in various petroleum fractions of different geological origin.[2, 4-7] Considerable insight has been gained from these findings.

ESR methods can be applied to reveal three of the most important types of information concerning the vanadium bonding of petroleum:

1. Nature of ligand types: ESR correlation techniques can pinpoint the composition of heteroatoms coordinated to vanadium; for example, possible combinations

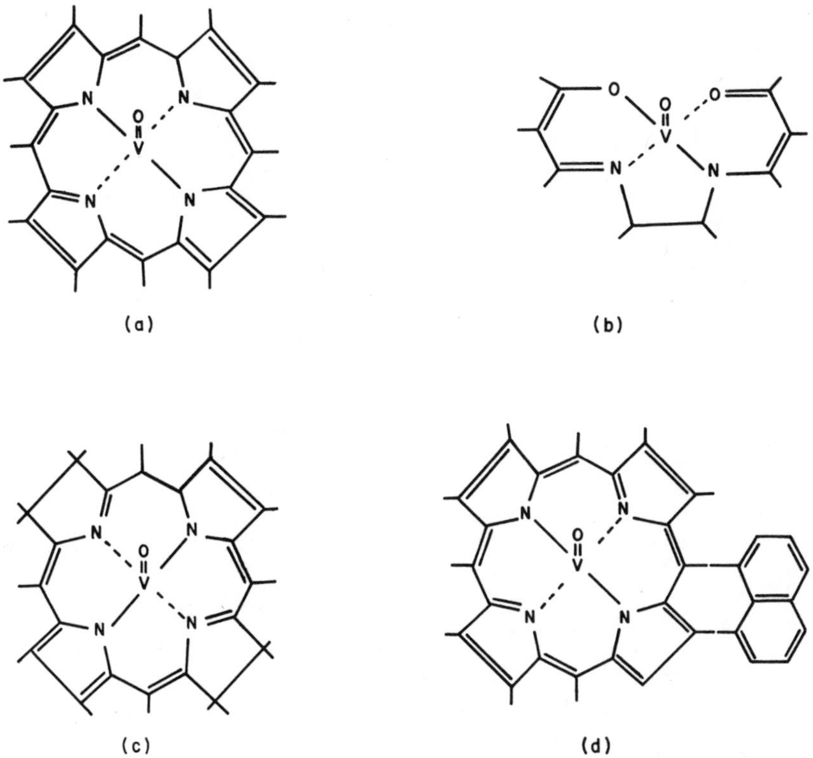

(a)   (b)

(c)   (d)

Figure 10.1.   Key types of vanadium complexes in petroleum:
(a) porphyrins, (b) mixed ligands such as N_2O_2
type, the β-ketimines, (c) pseudoaromatic
pheophorbides such as a bacterio-chlorophyll
(the outer conjugation is interrupted, but
still belong to the diaza-18-annulene system),
and (d) highly aromatic porphins such as the
dehydrogenated product of m-α-naphylporphyrin,
which was identified in Nonesuch shale. Type
(a) is found in all petroleum. Types (b), (c)
and (d) are commonly referred to as the nonporphyrin type of vanadium

of four ligands may include: all nitrogen (N_4),
half nitrogen, half sulfur (N_2S_2), or "odd" nitrogen
compositions (NX_3 or N_3X, where X is O or S) (Figure 10.2).

2. Distortion of complexes: Usually vanadyl complexes
    are square pyramidal; the four donor atoms are
    coplanar in the basal planes. Reduction of symmetry
    from the axial to rhombohedral can be detected.

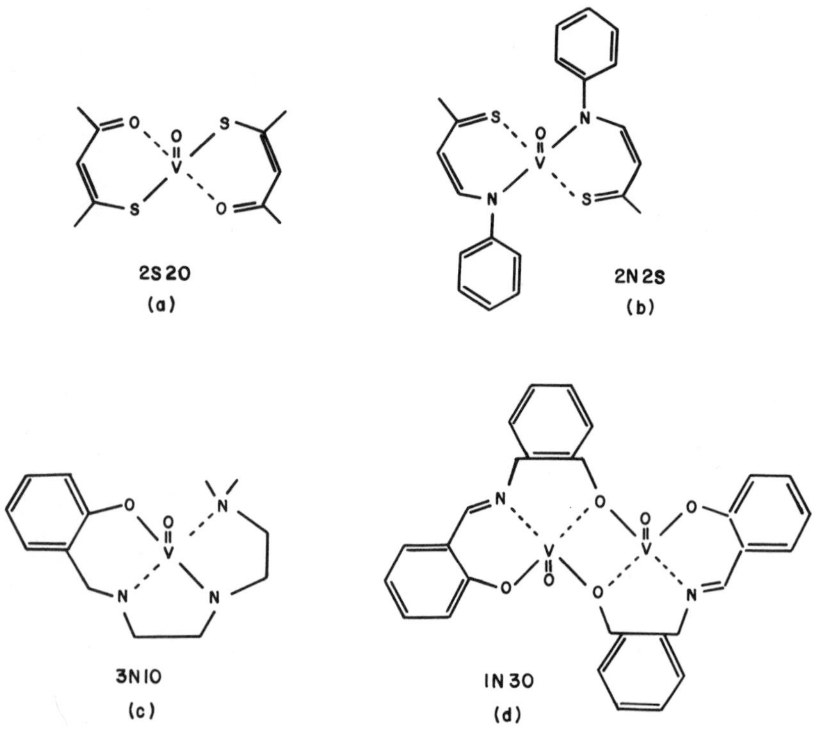

Figure 10.2.   Vanadyl quadrivalent ligand types: (a) 2S2O
type, (b) 2N2S type, (c) 3N1O type, and (d)
1N3O type.   Types (c) and (d) belong to the
odd nitrogen class.   For the 2N2O type, see
Figure 10.1 (b).

3.   Association with π-systems: The influence of donor-
acceptor properties of a large aromatic π-system
exerted on the vanadium system will change the
vertical distance of vanadium atoms to the approximate ligand plane of the heterocyclic atoms (Figure
10.3).

Information obtained from the above studies
will help not only to elucidate the structures of
bitumens, kerogens and other naturally occurring
substances but also to clarify the mechanisms of
biogenesis, transformation and metamorphism of all
fossil remains existing in nature.   A better understanding of the interactions of vanadium complexes

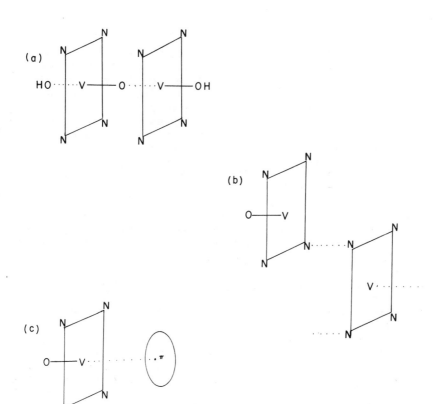

*Figure 10.3. Possible structure of vanadyl association:
(a) V-O•••V type, (b) O-V•••N type, and (c)
O-V•••π type. For type (c), the π-system could
be either from the nonporphyrin highly aromatic
porphins (Figure 10.1, d), or from the host
aromatic sheets of asphaltic system.*

in these bituminous materials will allow the improved control of petroleum refining, processing,
cracking, and conversion processes as well as the
development of more effective methods for desulfurization and denitrification of fuels for environmental pollutants abatement.

EXPERIMENTAL

The isolation of vanadium complexes from fossil remains adapted the protocol established for the Lunar fines, which was developed by this investigator.[8-10] The conventional scheme for further separation by column and gel-permeation chromatography of these grossly separated species was attempted. A successive solvent elution with hexane-benzene-ether-methanol series of increasing polarity, with either neutral aluminum or the Sephadex as the stationary phase, was followed.

Each individual fraction was examined by ESR for the isotropic region of the g_O vs. a_O. From the correlation of g_O and a_O, comparative information concerning the ligand heteroatoms was obtained.

Samples also were examined for the presence of an anisotropic spectra to detect any splitting of the $g_\perp$ into g_x and g_y. Degrees of reduction of symmetry were evaluated for a number of low-symmetry complexes. Parameters such as Δ, spacing differences, and ρ, axial to the equatorial crystal field were obtained.[6, 36]

The nature of nitrogen superhyperfine (shf) splittings of the vanadyl groupings within these samples was examined. In this manner, the interaction caused by the donor-acceptor associations is accessed and the environment of the ligand atoms is revealed (such as equivalence of donor atoms), sixth vacant position coordination, and large π-π association). This is especially important for the aryl-substituted or π-fused porphyrins.

DATA TREATMENT

ESR studies can yield two types of information on bitumens and related substances: (a) free radical absorption, and (b) absorptions due to naturally occurring vanadium and synthetic vanadium chelate model systems. The studies prior to 1968 have been summarized in a review by this investigator.[6] Recently, considerable ESR work concerning fossil remains has been carried out by this investigator and his associates: g-value correlation,[7] temperature-dependence studies,[11] stability,[12] transformation,[13] nitrogen superhyperfine splittings,[14] ligand types and correlations,[15] anisotropy-isotropy,[5] model chelate systems,[16-20] spectral calculation,[21] synthesis,[22] and enhancement and separation of vanadium signals.[23]

The ligand correlation of the vanadyl environment was reexamined.[24] The basis for the analysis is the correlation of g_0, center of signal, and a_0, hyperfine constant of an isotropic spectra:[15]

$$-a_0 = P \, (g_e - g_0) + PK$$

where P is the direct dipolar term and K is the Fermi contact term. For a series of vanadyl complexes $P = 136$ G and the variation of K is small: Thus, when a_0 *vs.* g_0 was plotted, an approximately linear relation was obtained for ligands made of varying compositions (Figure 10.4). The correlation becomes a more useful method when recent data including sulfur is available (Figure 10.5).[25, 26] This method is applicable to any sharp fractions of petroleum.

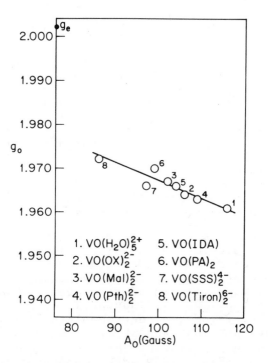

Figure 10.4. An approximately linear relation of vanadyl complexes of varying compositions.

The problem concerning small distortion or deviation from axial symmetry has been elucidated.[20] Coordination of the sixth vacant position was also

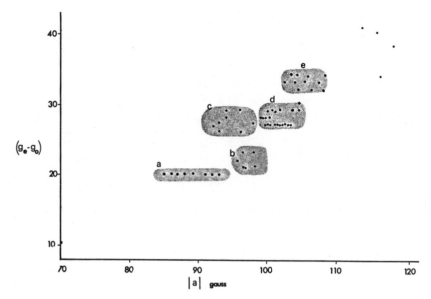

Figure 10.5. *Correlation of ligand types of vanadyl complexes with ESR determinations (a) S_4, (b) N_4, (c) O_2S_2, (d) N_2O_2, and (e) O_4.*

studied.[16] Calculation of spectra was developed[22] in order to detect small differences of the x,y splitting.

The superhyperfine (shf) lines due to the coupling of unpaired spins to 14_N nuclei in both native asphaltenes derived from Boscan and Bachaquero crude oils were observed.[6] These patterns usually consisted of nine equally spaced lines with intensity ratios of 1:4:10:16:19:16:10:4:1 yielding a^N of 2.75 G. In addition, shf splittings were observed[14] when an oxovanadium (IV) macrocycle was added to a host consisting of a fused-ring aromatic system[6] (*e.g.*, perylene, 9,10-phenanthroline, phenanthrene polymer, and vanadium-free asphaltene).

Other than this recent finding, many researchers have studied different vanadyl complexes with ESR and have failed to observe positively nitrogen shf structures except one porphyrin (tetraphenylporphin) under one set of special conditions (low temperature, chloroform or CS_2 glasses).[27] Nitrogen shf splittings obtained at these conditions have not been explained.

DISCUSSION

The ligands of biotic vanadyl complexes can take on a number of different forms, e.g., hemovanadium is N_2O_2 (2 nitrogens and 2 oxygens), whereas cellular vanadium takes the N_4 form, although all the N's are not necessarily identical. After these biotic complexes had undergone fossilization, those remaining were the more thermodynamically stable forms. Usually the resulting mixture consists of heterogeneous components. Literature reported to date[2,28] indicates that the vanadium compounds in fossil fuels are not all homogeneous; they consist mostly of porphyrins (including pheophorbides of cycloalkanoporphin), arylporphins (rhodoporphyrins), complexes of interrupted conjugation (pseudoaromatic ligands), and mixed ligands (nonporphyrins).

Often in sediments vanadyl groups are associated as ligands with the etioporphyrins although virtually nothing is known about the nature of the remaining ligands. This is due to the fact that a limited number of vanadyl etioporphyrin species are loosely bound to the matrix of fossil remains (if the sample is washed with methanol, the etiospecies are extracted). Therefore, work reported in this area is generally concerned only with the etioporphyrin types contained in the fossil remains. In this regard, it may be stated that vanadium, as an element, is easy to detect, but that the structure of the vanadium sites in fossil remains is difficult to elucidate.

With drastic variations in sediments, composition as well as types of vanadium complexes vary. Even within a series of samples there is a noticeable variation of the vanadium complexes. For example, as the age or the depth of the fossil remains increases,[13,12] a transformation of the porphyrin types from *philo* to *etio* takes place. This is merely one example of the large variety of reactions and transformations occurring in petroleum as well as other bituminous substances. The study of such effects would reveal much knowledge concerning the migration, maturation, and transformation of a diversity of fossil remains.

All known X-ray diffraction data indicate that in penta-coordinated pyramidal vanadium chelates the vanadium atom lies out of the basal plane of the ligand macrocycle, *i.e.*, the plane of the four pyrrole nitrogen atoms. The distance from vanadium

atom to the basal plane ($d_{V \to p}$) for vanadyl deoxy-
philloerythroetioporphyrin (DPEP) is 0.53 A[29] and
for vanadyl β-ketimines[20] is 0.47 A.   Recently, however, one of the preparations of a vanadyl β-
ketimine, N,N'-propylene-bis-(salicylaldiminato)
oxovanadium, or VO (salpn), was reported[30] to have
an unusual coloration and a $d_{V \to p}$ of 0.31 A, which
is shorter than most vanadyl complexes.[31]   This
anomaly could be attributed to the ligation of
···V - O - V···bonding.[3, 31]   Similarly, in VO $(H_2O)_4$
SO_4 ·H_2O, the water molecule in the sixth position
(normally vacant) of the square pyramid causes $d_{V \to p}$
to be reduced to 0.29 A.[32]   From these examples it
appears that the $d_{V \to p}$ of a vanadyl chelate depends
on the electronic interaction of the microenviron-
ments of the sixth vacant axial position (Figure
10.6).

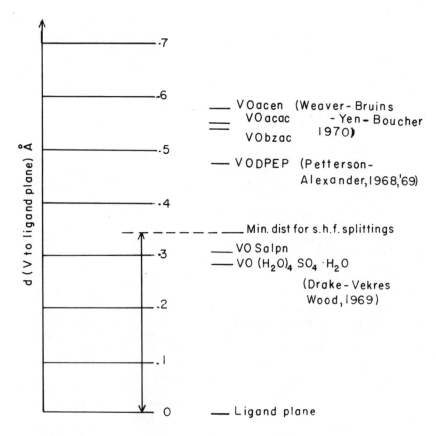

Figure 10.6.   X-ray data for the V to ligand plane distance.

A molecular orbital calculation[33] (Figure 10.7) has shown that the antibonding coefficients, $(\beta_1^*)^2$ and $(e\pi^*)^2$, of vanadyl porphyrins vary as a function of $d_{V \to p}$ (Figure 10.8). Since the spin-orbital coupling constant ξ and the anisotropic $g_\perp$ and $g_\parallel$ tensors for a given vanadyl chelate are constant, and from the first order approximation,

$$\Delta E_1 = 8\xi(\beta_1^*)^2 (\beta_2^*)^2/(g_e - g_\parallel)$$

$$\Delta E_2 = 2\xi(\beta_2^*)^2 (e\pi^*)^2/(g_e - g_\perp)$$

thus, the d-d transitions, ΔE_1 and ΔE_2, are also functions of $d_{V \to p}$ (in this case, where $(\beta 2^*)^2$ is constant through the range studied. As the vanadium atom approaches the plane of the macrocycle, the transition due to π-orbital in-plane will increase, whereas that due to the π-orbital out-of-

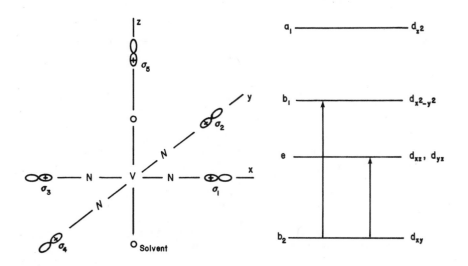

$$\Psi(b_2) = \beta_2(d_{xy}) + \tfrac{1}{2}\beta_2'(P_{y_1} + P_{x_2} - P_{y_3} - P_{x_4})$$

$$\Psi(b_1) = \beta_1(d_{x^2-y^2}) + \tfrac{1}{2}\beta_1'(\sigma_1 - \sigma_2 + \sigma_3 - \sigma_4)$$

$$\Psi(e_\pi) = \epsilon_\pi d_{xz} + \epsilon_\pi' P_{x_5} + \tfrac{1}{\sqrt{2}}\epsilon_\pi''(P_{21} - P_{22})$$

$$\Psi(a_1) = a_1(d_{z^2} + s_0) + a_1'\sigma_5$$

Figure 10.7. Molecular orbital scheme of vanadyl complexes.

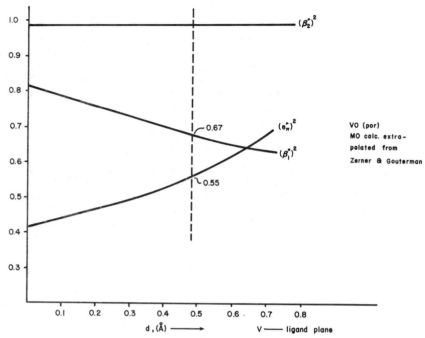

Figure 10.8.   Variations of antibonding coefficients with V
to ligand plane distance.

plane will decrease. For porphyrins, the strong
π-π transition will mask the weak d-d transitions
and consequently make the latter ones difficult
to observe directly.

At this stage, the unknown factor is whether
ΔE_1 and ΔE_2 can be influenced by the aromatic system axially at the sixth position of porphyrins.
However, it is known that porphyrins and phthalocy-
amines behave like azaannulenes and azaarenes, in
which the π-π overlap is significantly large in the
presence of other fused-ring π-systems. Even for
the mesomorphic asphaltene molecules, an extensive
charge-transfer process among the large condensed-
ring system was observed.[11, 34, 35] Apparently, a
large π-system close to the back side of the chelate
will exert an influence on ΔE_1 and ΔE_2, but this
requires further experimental support. In this con-
nection it is worthy to note that for vanadyl β-
diketones both ΔE_1 and ΔE_2 are extremely solvent-
dependent due to back coordination (Table 10.1)[37]

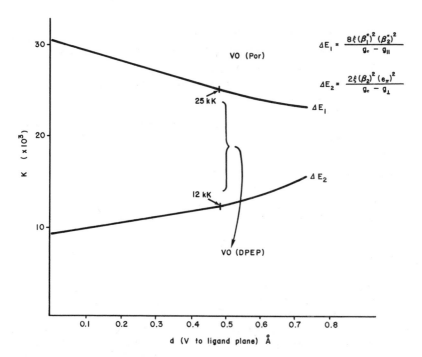

$$\Delta E_1 = \frac{8\xi(\beta_1^*)^2(\beta_2^*)^2}{g_e - g_{\parallel}}$$

$$\Delta E_2 = \frac{2\xi(\beta_2^*)^2(e_\pi)^2}{g_e - g_{\perp}}$$

Figure 10.9.   Relation of d-d transition energies to V to
ligand plane distance.

Table 10.1

Experimental Bank Maxima for Vanadyl β-Diketones (kk)

Sixth Ligand	I	II	III	DII.I
Methanol	13.03	17.45	25.63	4.42
n-Propylamine	12.94	17.30	24.69	4.36
Ethanol	12.98	17.24	(24.99)	4.18
Pyridine	12.99	17.22	23.80	4.21
Dimethylsulfooxide	12.85	17.03	24.99	4.18
n-Butanol	13.2	17.3	25.4	4.1
Dioxane	13.83	17.85	---	4.02
Piperidine	13.28	17.30	---	4.02
Dimethylformamide	13.03	16.97	25.12	3.94
n-Pentanol	13.3	17.1	25.2	3.8
Formic acid	13.28	16.52	---	3.25
Tetrahydrofuran	13.66	16.80	---	3.14
Acetonitrile	14.20	16.80	25.80	2.60
Nitromethane	14.57	16.83	(25.37)	2.26
Nitrobenzene	14.71	16.66	---	1.95
Chloroform	14.92	16.86	25.97	1.94
Benzene	15.31	16.89	25.70	1.58
m-Tylene	15.22	16.72	25.31	1.50
Carbon tetrachloride	15.79	16.86	25.97	1.07

On the basis of the above arguments, it is im-
perative that the following questions be answered:

1. Do aromatic hosts shorten $d_{v \to p}$?
2. Can one control the nature of the hosts to obtain
   desirable d-d transitions and therefore a desirable
   $d_{v \to p}$?
3. What is the maximun $d_{v \to p}$ that will permit the observation of nitrogen shf?

ACKNOWLEDGMENT

The financial support provided by ACS-PRF grant-in-aid
through No. 6272-AC2 is gratefully acknowledged.

REFERENCES

1. Yen, T. F. in *Trace Substances in Environmental Health,*
    vol. 6, D. D. Hemphill, Ed. (Columbia, Missouri:
    Univ. of Missouri, 1973) pp. 347-353.
2. Yen, T. E., L. J. Boucher, J. P. Dickie, E. C. Tynan
    and G. B. Vaughan. *J. Inst. Petroleum, 55, 87* (1969).
3. Rho, J. H., A. J. Bauman, H. G. Boetteger, and T. F. Yen.
    Space Life Sci., 4, 69 (1973).
4. Yen, T. F., J. G. Erdman and A. J. Saraceno. *Anal. Chem.,
    34, 694* (1962).
5. Tynan, E. D. and T. F. Yen. *Fuel* (London), *43, 191*
    (1969).
6. Yen, T. F., E. C. Tynan, G. B. Vaughan and L. J. Boucher
    in *Spectrometry of Fuels,* R. A. Friedel, Ed. (New York:
    Plenum Press, 1970) Chapter 14, pp. 187-201.
7. Yen, T. F. and S. Sprang. Preprints, *Div. of Petrol.
    Chem., ACS, 15,* No. 3, A65 (1970).
8. Rho, J. H., A. J. Bauman, T. F. Yen and J. Bonner.
    Science, 167, 754 (1970).
9. Rho, J. H., A. J. Bauman, T. F. Yen and J. Bonner.
    Proc. Apollo 11 Lunar Sci. Conf., vol. 2, A. A. Levinson,
    Ed. (New York: Pergamon Press, 1970) pp. 1929-1932.
10. Rho, J. H., A. J. Bauman, J. F. Bonner and T. F. Yen,
    in *Proc. 2nd Lunar Sci. Conf.,* vol. 2, A. A. Levinson,
    Ed. (Cambridge, Massachusetts: MIT Press, 1971) pp.
    1875-1878.
11. Yen, T. F. and D. K. Young. *Carbon 11,* 33 (1973).
12. Yen, T. F. in *Chemistry in Space Research,* R. F. Landel
    and A. Rembaum, Eds. (New York: Elsivier Publishing Co.,
    1972) pp. 105-153.
13. Yen, T. F. and S. K. Silverman. Preprints, *Div. of
    Petrol Chem., ACS, 14,* No. 3, E32 (1969).

14. Yen, T. F. *Naturwis*, *58*, 267 (1971).
15. Boucher, L. J., E. C. Tynan and T. F. Yen in *Electron Spin Resonance of Metal Complexes*, T. F. Yen, Ed. (New York: Plenum Press, 1969) pp. 111-130.
16. Boucher, L. J., E. C. Tynan and T. F. Yen. *Inorg. Chem.*, *7*, 731 (1968).
17. Boucher, L. J. and T. F. Yen. *Inorg. Chem.*, *7*, 2665 (1968).
18. Vaughan, G. B., E. C. Tynan and T. F. Yen. *Chem. Geol.*, *6*, 203 (1970).
19. Boucher, L. J. and T. F. Yen. *Inorg. Chem.*, *8*, 689 (1969).
20. Boucher, L. J., D. E. Bruins, T. F. Yen and D. L. Weaver. *J. Chem. Cos. D Chem. Comm.*, *7*, 363 (1969).
21. Makey, J. H., M. Kopp, E. C. Tynan and T. F. Yen in *Electron Spin Resonance of Metal Complexes*, T. F. Yen, Ed. (New York: Plenum Press, 1969) pp. 111-130.
22. Tynan, E. C. and T. F. Yen. *J. Magnetic Resonance*, *3*, 327 (1970).
23. Beasely, J. E., R. L. Anderson, J. P. Dickie, F. R. Dollish, E. C. Tynan and T. F. Yen. *Spectros. Lett.*, *2(5)*, 149 (1969).
24. Yen, T. F., L. J. Boucher, J. P. Dickie, E. C. Tynan and G. B. Vaughan. Preprints, *Div. of Petrol. Chem.*, *ACS*, *13*, No. 1, 59 (1968).
25. Yen, T. F. "Electron Spin Resonance of Vanadyl Porphyrins and other Chelates," Gordon Research Conferences on Geochemistry, Plymouth, New Hampshire, Aug. 31, 1970.
26. Dickson, F. E., C. J. Kunesh, E. L. McGinnis and L. Petrakis. Preprints, *Div. of Petrol. Chem.*, *ACS*, *16*, No. 1, A37 (1971).
27. Kivelson, D. and S. K. Lee. *J. Chem. Phys.*, *41*, 1896 (1964).
28. Baker, E. W., T. F. Yen, J. P. Dickie, R. E. Rhodes and L. F. Clark. *J. Amer. Chem. Soc.*, *89*, 3631 (1967).
29. Peterson, R. C. and L. E. Alexander. *J. Amer. Chem. Soc.*, *90*, 3873 (1968).
30. Mathew, M., A. J. Corty and G. J. Palenik. *J. Amer. Chem. Soc.*, *92*, 3197 (1970).
31. Bruins, D. E. and D. C. Weaver. *Inorg. Chem.*, *9*, 139 (1970).
32. Ballhausen, C. J., Djurinskij, and K. L. Watson. *J. Amer. Chem. Soc.*, *90*, 3305 (1968).
33. Zerner, M. and M. Gouterman. *Inorg. Chem.*, *5*, 1699 (1966).
34. Yen. T. F. Preprints, *Div. Fuel Chem.*, *ACS*, *15*, 93 (1971).
35. Sill, G. A. and T. F. Yen. *Fuel* (London), *43*, 61 (1969).
36. Kuska, H. A. and M. T. Rogers. *Inorg. Chem.* *5*, 313 (1966).
37. Selbin, J. *Chem. Rev.*, *65*, 153 (1965).

CHAPTER 11

OXIDATIVE DEMETALLATION OF OXOVANADIUM(IV)
PORPHYRINS

James M. Sugihara, J. F. Branthaver
and K. W. Willcox
Department of Chemistry, North Dakota State University
Fargo, North Dakota 58102

The presence of vanadium in petroleum crudes
is well documented. Much analytical data[1-3] have
been compiled because the metal brings about unfavor-
able effects in *poisoning cracking and other catalysts*
in the standard refinery operations. Furthermore,
presence of vanadium in residual fuels at high
levels results in damage to furnace components.
    Vanadium is found as a chelate of porphyrins
and in other forms, presently ill-defined. Of these
metal-containing compounds, only the metalloporphyrins
are volatile. Thus these species are potentially
more deleterious than the nonporphyrin vanadium compounds.
    A variety of methods has been established to
effect metal removal from vanadium porphyrins. For
the most part the procedures described require the
use of strong acids and are applied as a means of
determining porphyrin content. The common reagents
are hydrogen bromide in acetic acid,[4] hydrogen
bromide in formic acid,[5] sulfuric acid,[6] methanesulfonic acid,[7] and hydrogen fluoride.[8] Unfortu-
nately in all instances these reagents cause marked
changes to petroleum materials and thus are not satisfactory for purposes of metal removal. A reductive
procedure in a basic medium[9] also brings about deep-
seated changes to a petroleum sample.[10]
    In some recent studies, certain oxidative re-
agents were found to effect rapid demetalation of
vanadium porphyrins under mild conditions without

183

apparent serious alteration to the rest of the petroleum materials. These observations and those on synthetic vanadyl porphyrins are described herein.

EXPERIMENTAL

Most of the studies were carried out on Boscan crude and fractions derived from it because these materials are rich in vanadium and metalloporphyrins. Boscan asphaltenes were obtained by pentane precipitation. The asphaltenes were further separated by dissolving in benzene and subjecting to gel permeation chromatography, using an 80 Å gel of polystyrene cross-linked with divinylbenzene, and a column of 120 cm height and 2.1 cm radius. The charge consisted of 3.0 g asphaltenes in 30 ml benzene, with the elution rate maintained at 6 ml/min. Breakthrough occurred at 450 ml, the first 120 ml of dark-colored eluate was collected as Fraction I, the next 180 ml of eluate was not used in this study, and the final 600 ml was collected as Fraction II. Metalloporphyrin[11] and vanadium[12] contents of these materials were determined; data are given in Table 11.1.

Table 11.1

Vanadium and Metalloporphyrin Contents
in Boscan Crude and its Fractions

	V (μ Moles/g	Metalloporphyrin (μ Moles/g)
Boscan crude	22.3	10.4
Asphaltenes	70.0	29.1
Fraction I	75.0	15.0
Fraction II	72.0	79.0

Reagents found to have varying capabilities in effecting vanadium removal are given in a descending order of reactivity: chlorine $\simeq$ sulfuryl chloride > t-butyl hypochlorite > dinitrogen tetroxide > t-butyl hydroperoxide $\simeq$ benzoyl peroxide $\simeq$ azobisisobutyronitrile >

bromine.   In a typical run 1.25 g of petroleum material
was dissolved in 50 ml of methylene chloride (or
toluene) and placed in a 500-ml three-necked flask
equipped with a dropping funnel, stirrer, and drying
tube.   The contents were cooled to -78°C when the
reactive reagents, chlorine and sulfuryl chloride,
were used; in other instances 0°C, ambient temperatures, and temperatures up to 110°C were maintained
to effect reactions.   The crude oil was also treated
neat with no added solvent.   Small amounts of reagent,
usually dissolved in methylene chloride, were added
into the stirred mixture.   A minimum of 10 moles of
reagent per mole of metalloporphyrin was required.
Reaction was allowed to continue for varying periods
of time, as short as one minute for chlorine and
sulfuryl chloride.   Reactions using chlorine or sulfuryl chloride were quenched by addition of phenol or
ammonium hydroxide.   Other reactions were terminated
by cooling the reaction mixture.   The reaction mixture was extracted with 5% aqueous sodium hydroxide
and then with water.   Vanadium and metalloporphyrin
determinations were then made.   Some of the data
are given in Table 11.2.
      Etioporphyrin I was synthesized[13] and converted[14]
into its oxovanadium(IV) complex.   Tetraphenylporphine was made by condensation of pyrrole and ben-
zaldehyde,[15] octaethylporphine was prepared,[16] and
both were converted[17] into their oxovanadium(IV)
complexes.   Reactions on the synthetic porphyrins
and their vanadyl complexes were carried out in much
the same fashion as the petroleum fractions.

RESULTS

      Some of the data obtained in observing vanadium
removal and metalloporphyrin destruction are given in
Table 11.2.   Chlorine and sulfuryl chloride were very
effective under surprisingly gentle reaction conditions.
The experiment using Fraction II is particularly note-
worthy because this material has all of its metal
content accountable as metalloporphyrins, 72.0 μM/g
vanadium and 7.0 μM/g nickel.   Following reaction
with chlorine, vanadium content was decreased from
79.0 μM/g to 5 μM/g, and metalloporphyrin content
was reduced to a nondetectable point.   Nonporphyrin
metal contents were not reduced nearly to the same
extent, though they were significantly altered.
      In order to interpret better the nature of the
reactions involved in these reactions involving vanadium removal and metalloporphyrin destruction, extensive studies have been made on synthetic porphyrins

Table 11.2

Vanadium and Metalloporphyrin Contents
Following Reactions

Material	Reagent	Conditions	V ($\mu M/g$)	Metalloporphyrin ($\mu M/g$)
Asphaltenes in toluene	Chlorine	-78°C, 1 hr.	35	0
Fraction I in toluene	Chlorine	-78°C, 1 hr.	39	0
Fraction II in methylene chloride	Chlorine	-78°C, 1 min.	5	0
Boscan crude in methylene chloride	Chlorine in methylene chloride	-78°C, 1 min.	--	0
Boscan crude in methylene chloride	Sulfuryl chloride in methylene chloride	-78°C, 5 min.	9	0
Boscan crude in methylene chloride	Sulfuryl chloride in methylene chloride	-78°C, 1 min.	14	~1
Boscan crude	Chlorine in methylene chloride	-78°C, 1 min.	--	< 1
Boscan crude	t-Butyl hydroperoxide	-110°C, 22 hrs.	--	4.7

and their oxovanadium(IV) complexes. Interestingly,
all the synthetic vanadyl porphyrins were found to
be less reactive than the petroporphyrins in Boscan
crude. The vanadyl complex of any given porphyrin
was observed to be more reactive than the parent porphyrin, which in turn was more reactive than the
diprotonated species. Vanadyl tetraphenylporphine
was determined to be much less reactive than vanadyl
etioporphyrin I or vanadyl octaethylporphine.

Light accelerated all reactions, but the order
found for reactivity in the dark, chlorine ≈ sulfuryl
chloride > t-butyl hypochlorite > dinitrogen tetroxide >
t-butyl hydroperoxide ≈ benzoyl peroxide ≈ azobisisobutyronitrile > bromine, was still maintained. When
light entering the reaction mixture was filtered
through a solution of vanadyl etioporphyrin I, the
resulting radiation produced no acceleration.

The reactivity of fluorine toward synthetic vanadyl porphyrins, Boscan crude, and fractions derived
therefrom was surprisingly benign.[18] The halogen was
actually less reactive than chlorine toward metalloporphyrins. Though the crude and its fractions did
react with the formation of hydrogen fluoride, the
processes did not exhibit explosive violence at ambient
and even higher temperatures.

Several approaches were applied in attempting to
establish the nature of the products formed in these
reactions. When synthetic vanadyl porphyrins were
treated with chlorine in the presence of hydrogen
chloride, the diprotonated porphyrin could not be
trapped, although reactivity of the metalloporphyrin
is appreciably greater than that of the diprotonated
porphyrin. Product analysis was made on a run involving 300 mg of vanadyl octaethylporphine and 5 molar
equivalents of sulfuryl chloride in methylene chloride in the dark at −78°C. Reaction was quenched with
ammonium hydroxide. The resulting material was washed
with water, dried, and then chromatographed on
Silicar-CC-7,[19] using benzene and benzene-methanol as
eluants. Unreacted vanadyl octaethylporphine was
recovered from the benzene eluant as well as a green
fraction, which exhibited an absorbance near 450 nm
and contained chlorine and vanadium at a level that
would be expected for chlorinated vanadyl octaethylporphine. The green fraction from the benzene-methanol
eluate also exhibited an absorbance near 450 nm and
contained very little vanadium. The metal content was
low enough that a proton nuclear magnetic resonance
spectrum was obtainable. This spectrum demonstrated
the disappearance of methine, a decrease in methylene,
and essentially no change in methyl protons.

DISCUSSION

The unexpected order of reactivity of the halogens, namely chlorine > fluorine > bromine, suggests
that an initial interaction of vanadyl porphyrins
and these reagents occurs, which influences the course
of the process. This interaction may involve formation
of a π-complex (charge-transfer complex) with the
metalloporphyrin functioning as an electron donor to
the electron-deficient halogen molecule. Of the common halogens, iodine would be the best acceptor,
presumably because of its greater ease of polariza-
bility, and fluorine a poor acceptor. Loss of an
electron from the metalloporphyrin would lead to a
cation-radical. Cation-radicals have been shown to
be intermediates in reactions of metalloporphyrins with
benzoyl peroxide,[20, 21] bromine,[22-24] hydrogen peroxide,[25]
or by electrochemical oxidation.[28-32] Indeed, cation-
radicals have been proposed[33-35] to be involved in
electrophilic aromatic substitution reactions.

The observation that light accelerates these
reactions can be rationalized in the framework of the
above interpretation. Since light passed through a
solution of vanadyl etioporphyrin I without effect on
the system, it would appear that photochemically excited species of metalloporphyrins are responsible for
the faster reactions as a consequence of their enhanced
capability of interaction with halogens and the other
reagents.

The metalloporphyrin cation-radical would likely
react with the radical derived from the reagent, with
substitution occurring at a meso-position, leading to
the formation of a resonance-stabilized carbonium ion
of the metalloporphyrin. That substitution occurs in
the meso-positions is suggested by the observation that
one of the major fractions obtained from the reaction
of vanadyl octaethylporphine with sulfuryl chloride
contained no methine protons, as demonstrated by the
nmr spectrum. The site of substitution being meso is
further supported by the fact that reactions of
metalloporphyrins with peroxides[20, 21, 25-27] and bromine[22-24] provide meso-substituted products. Electrophilic reagents in general react with porphyrins and
metalloporphyrins with meso-substitution.[36-42] The
possibility that some type of direct attack occurred
on the central metal, leading to its ejection, seems
very unlikely since the protonated porphyrin, much
less reactive than the metalloporphyrin, could not be
trapped in the reaction mixture in the presence of
added hydrogen chloride.

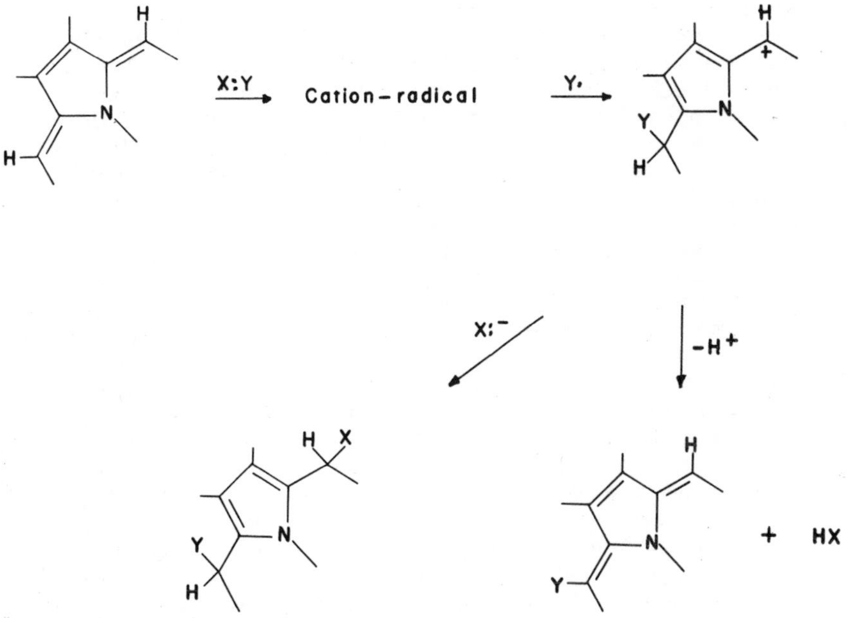

Figure 11.1

It would appear logical that the metalloporphyrin cation could then undergo reaction by two different paths, one involving proton ejection to yield the product of substitution and one involving combination with an anion to form the product of addition. Poly-nuclear aromatic hydrocarbons frequently undergo re-action with halogens to give products of both substi-tution and addition. The driving force leading to the product of substitution is restoration of reson-ance stabilization. The formation of the product of addition from the metalloporphyrin cation would yield a product that would have its macrocyclic, con-jugated structure disturbed. Such a species should be an appreciably poorer ligand towards the metal, allowing for the loss of vanadium by a displacement process. Thus the two fractions obtained in the reaction of vanadyl octaethylporphine with sulfuryl

chloride may be those products obtained by a substitution process retaining vanadium and those by an
addition mechanism with loss of vanadium.

A molecule formed by the substitution path could
undergo further reaction until all meso-positions
are substituted. The product obtained by addition
may undergo a variety of reactions beyond metal
loss, since the molecule is merely an alkylated pyrrole
and reactive to electrophilic reagent.

The greater lability toward vanadium removal and
porphyrin destruction for the vanadyl petroporphyrins
over the synthetic vanadyl porphyrins is a fortunate
circumstance. This difference can be rationalized
based upon structural differences of the porphyrins
involved. Phyllo-type petroporphyrins all contain a
cyclopentane ring, fused to one of the pyrrole rings
with the methine carbon one of the units of the carbo-
cyclic structure. Etio- and rhodo-type petroporphyrins
appear[43] to have alkyl substitution at one or more
methine positions, as based upon nmr spectral data.
Should all petroporphyrins have a carbon substituent
on one or more methine carbons, the carbonium ion
formed by the reaction would tend to have more charge
localization on those methine carbons. These ions
are a more stable species by an order of magnitude
over those porphyrins without methine substitution.
Such a carbonium ion would offer a better opportunity
for the addition path over the elimination path.

The fact that vanadyl tetraphenylporphine was
less reactive is not inconsistent with this rational-
ization although in this instance all four methine
carbons have phenyl substituents. It is likely that
the formation of the mono-substituted carbonium ion
would be inhibited because of steric interactions
at the methine position involved. However, a more
important consideration is that the benzene ring at-
tached to the methine carbon is not coplanar with
the macrocyclic porphyrin ring.[44] The extent of
deviation from planarity is enough that a positive
charge generated on the methine carbon would receive
no resonance stabilization but considerable inductive
destabilization as a consequence of the unsaturated
substituent.

The reactions, particularly of the chlorine-
containing reactants, offer a new approach to metal-
removal from crudes and heavy residues. High reactivity of metalloporphyrins was exhibited even at
low concentrations and at low temperatures. Changes
in the petroleum material are likely to be minimal.
Treatment of the crude or any of its fractions with

chlorine or sulfuryl chloride did not alter separation
characteristics on the gel permeation column, other
than effect loss of the color in that fraction,
normally heavily laden with metalloporphyrins.
The observation that metallopetroporphyrins
could be removed completely with retention of considerable nonporphyrin vanadium provides one further
piece of evidence that nonporphyrin vanadium does not
exist as metal chelated with high molecular weight
porphyrins. Indeed, this approach provides the means
of obtaining petroleum material containing nonporphyrin
vanadium, free from porphyrins.

ACKNOWLEDGMENT

Financial support from the American Petroleum Institute
(Research Project 60B) is gratefully acknowledged.

REFERENCES

1. Whisman, M. L., and F. D. Cotton. *Oil and Gas J.*
    (Dec. 27, 1971), p. 111.
2. Horr, C. A., A. T. Myers, P. J. Dunton, and H. J. Heyden.
    "Uranium and Other Metals in Crude Oils," U.S. Geological
    Survey Bulletin 1100 (Washington, D.C.: U.S. Government
    Printing Office, 1961).
3. Bonham, L. C. *Bull. Am. Assoc. Petroleum Geol., 40,*
    897 (1956).
4. Groennings, S. *Anal. Chem., 25,* 938 (1953).
5. Sugihara, J. M., and R. G. Garvey. *Anal. Chem. 36,*
    2374 (1964).
6. Barnes, J. W. and G. D. Dorough. *J. Amer. Chem. Soc.,
    72,* 4045 (1950).
7. Erdman, J. G. U.S. Patent 3,190,829 (1965).
8. Kimberlin, C. N., Jr., H. G. Ellert, C. E. Adams and
    G. P. Hamner. U.S. Patent 3,203,892 (1965).
9. Eisner, U., and M. J. C. Harding. *J. Chem. Soc.,* 4089
    (1964).
10. Sugihara, J. M., T. Okada and J. F. Branthaver. *J. Chem.
    Eng. Data, 10,* 190 (1965).
11. Bean, R. M. and J. M. Sugihara. *J. Chem. Eng. Data, 7,*
    269 (1962).
12. Bean, R. M. Ph.D. Thesis, University of Utah, Salt Lake
    City, Utah, 1961.
13. Rislove, D. J., A. T. O'Brien and J. M. Sugihara.
    J. Chem. Eng. Data, 13, 588 (1968).
14. DeRemer, C. M. M.S. Thesis, North Dakota State University,
    Fargo, N.D., 1973.

15. Adler, A. D., F. R. Longo, J. D. Finarelli, J. Goldmacher, J. Assour and L. Korsakoff. *J. Org. Chem.*, *32*, 476 (1967).
16. Samuels, E., R. Shuttleworth and T. S. Stevens. *J. Chem. Soc.*, *Sect. C*, 145 (1968).
17. Adler, A. D., F. R. Longo, F. Kampas and J. Kim. *J. Inorg. Nucl. Chem.*, *32*, 2443 (1970).
18. Kowanko, N., J. F. Branthaver and J. M. Sugihara, 1973, unpublished observations.
19. A product of Mallinckrodt Chemical Co.
20. Pedersen, C. J., *J. Org. Chem.*, *22*, 127 (1957).
21. Bonnett, R. and A. McDonagh. *Chem. Commun.*, 337 (1970).
22. Felton, R. H., D. Dolphin, D. C. Borg and J. Fajer. *J. Amer. Chem. Soc.*, *91*, 196 (1969).
23. Fajer, J., D. C. Borg, A. Forman, D. Dolphin and R. H. Felton. *J. Amer. Chem Soc.*, *92*, 3451 (1970).
24. Fuhrhop, J. H., P. Wasser, D. Riesner and D. Mauzerall. *J. Amer. Chem Soc.*, *94*, 7996 (1972).
25. Bonnett, R., M. J. Dimsdale and G. F. Stephenson, *J. Chem. Soc.*, *Sect. C*, 564 (1969).
26. Bonnett, R. and M. J. Dimsdale. *Tetrahedron Lett.*, 731 (1968).
27. Bonnett, R., and M. J. Dimsdale. *J. Chem Soc.*, *Perkin Trans. 1*, 2540 (1972).
28. Fuhrhop, J. H. and D. Mauzerall. *J. Amer. Chem Soc.*, *90*, 3875 (1968).
29. Fuhrhop, J. H. and D. Mauzerall. *J. Amer. Chem Soc.*, *91*, 4174 (1969).
30. Fuhrhop, J. H., K. M. Kadish and D. G. Davis. *J. Amer. Chem. Soc.*, *95*, 5140 (1973).
31. Wolberg, A. and J. Manassen. *J. Amer. Chem. Soc.*, *92*, 2982 (1970).
32. Ferguson, J. A., T. J. Meyer and D. G. Whitten. *Inorg. Chem.*, *11*, 2767 (1972).
33. Brown, R. D. *J. Chem. Soc.*, 2224, 2232 (1959).
34. Nagakura, S. *Tetrahedron*, *19*, *Suppl. 2*, 361 (1963).
35. Pedersen, E. B., T. E. Petersen, K. Torsell, and S. O. Lawesson. *Tetrahedron*, *29*, 579 (1973).
36. Bonnett, R. and G. F. Stephenson. *Proc. Chem. Soc.*, 291 (1964).
37. Bonnett, R. and G. F. Stephenson. *J. Org. Chem.*, *30*, 2791 (1965).
38. Bonnett, R., I. A. D. Gale and G. F. Stephenson. *J. Chem. Soc.*, *Sect. C*, 1600 (1966).
39. Bonnett, R., I. A. D. Gale and G. F. Stephenson. *J. Chem. Soc.*, *Sect. C*, 1168 (1967).
40. Paine, J. B., III and D. Dolphin. *J. Amer. Chem. Soc.*, *93*, 4080 (1971).
41. Johnson, A. W. and D. Oldfield. *Tetrahedron Lett.*, 1549 (1964).

42. Grigg, R., A. Sweeney and A. W. Johnson. *Chem. Commun.*, 1237 (1970).
43. Branthaver, J. F. and J. M. Sugihara. 1972, unpublished observations.
44. Hoard, J. L., M. J. Hamor and T. A. Hamor. *J. Amer. Chem. Soc.*, *85*, 2334 (1963).

CHAPTER 12

KINETICS STUDIES OF DEMETALLATION
OF METALLOPORPHYRINS BY ACIDS

Betsey Don and T. F. Yen
Departments of Chemistry and Chemical Engineering
University of Southern California
Los Angeles, California

The occurrence of porphyrin metallo complexes
in any ancient sediment can be taken as a biological marker, and their presence in meteorites,[1,2]
interstellar space[3] or lunar samples[4,5] would suggest the presence of life. Porphyrins present in
petroleum are not homogeneous[6,7,8] and vary greatly
depending on geological ages.[9] The diagenic process points to the low temperature biological origin
of petroleum.

The porphyrin metallo complexes in crude oils,
asphaltenes and other natural bitumens are chiefly
those of vanadium and nickel[10] although copper,
iron and even uranium have been suggested.[11] Recently in a Precambrian shale, porphins were found
to chelate with iron, zinc and copper in addition
to vanadium and nickel.[12] The origin of these complexes is still uncertain, although several theories
have been advanced.[13] Some of these theories could
be verified or possibly even disproved if the porphyrin type bound to each metal was known. Furthermore, since these heavy metals are harmful to both
health and catalysts, a systematic study of demetallation of metalloporphyrins should prove useful.

The demetallation of metalloporphyrins by acid
is a reversible one and can be represented in a
generalized form by the following equation:

$$PM + HX \rightleftharpoons PH + MX$$

The kinetics of this can be conceivably described
as:

195

$$\frac{d\ [PM]}{dt} = k\ PM^m\ [HX]^n$$

where k is the demetallation reaction rate. Caughey and Corwin[14] have found that the demetallation of Cu-Etio II (copper etioporphyrin II) with sulfuric acid in acetic acid follows the above equation with m=2 and n=5. In the presence of excess acid, the right to left reaction is suppressed; in as much as [HX] remains essentially constant throughout the reaction, the kinetic equation can be reduced to the following form:

$$\frac{d\ [PM]}{dt} = k_{obs}[PM]^m$$

where $k_{obs} = k[HX]^n$. The values of n and k are obtained from the following relationship:

$$\ln k_{obs} = \ln k + n \ln [HX]$$

The present study was made in a nonaqueous medium using glacial acetic acid, which provided a good polar medium. Work was done on two porphyrins, Cu-TPP (copper ms-tetraphenylporphyrin) and Cu-Etio I (copper etioporphyrin I), with three acids-- hydrochloric, sulfuric, and methanesulfonic.

Cu-Etio I was found to be more susceptible to demetallation than Cu-TPP. In a 0.4 N acid medium, Cu-Etio I underwent an appreciable conversion over 2.5 hr, whereas Cu-TPP remained essentially unchanged. However, work was concentrated on Cu-TPP due to its greater availability.

Hydrochloric acid proved to be inferior to the other two acids used. Not only was its solubility in acetic acid low (continuous bubbling of HCl gas through glacial HOAc for 5 hr gave a solution of acid strength of 0.951 N), offering a limited acid strength range, but also it was not stable in solution, rendering difficulty in controlling the acid strength. Sulfuric acid provided a limited range due to its immiscibility with benzene at an acid concentration above 4 N. Methanesulfonic acid offered the widest workable range, was easiest to handle, and proved to be the most useful demetallation reagent.

All absorbance data were obtained on a Beckman DU-2 spectrophotometer, and quartz cells of 1 cm path length were used. The apparatus used was essentially a three-necked round-bottomed flask fitted with an electric stirrer and a thermometer; the third neck was left unoccupied for addition of

reagents. Most of the flask was submerged in a
constant-temperature bath maintained at a temperature of 24± 0.5°C.

Acetic acid and metalloporphyrin solutions
were added to the flask in quantities to give final
porphyrin concentration of approximately 10^{-4} M.
It was found that the reaction rate was not very
sensitive to minor changes in porphyrin concentration. The mixture was stirred for several minutes
to insure complete mixing, and a sample was removed
for readings of initial porphyrin concentrations.
When the temperature of the solution in the flask
had stabilized, acid was delivered through a volu-
metric pipette to make up the final volume, and at
the same time the timer was started. Aliquots
were removed at different noted times from the reaction flask and quenched with a 2:1:1 pyridine:
ethanol:water mixture. This specific combination
was chosen to insure a phase system in which ethanol
would hold benzene in solution, and water would dis-
solve any acid-pyridine salts. The quantities used
were 1 ml sample aliquots to 4 ml quenching solution to insure the presence of excess base and the
dilution of porphyrin to give a convenient absorb-
ence reading range (below 1). The final total
porphyrin concentration (metallo + free) was kept
in the neighborhood of 2 x 10^{-5} M.

Assuming the reaction to be 1st and 2nd order
respectively, the following relationships can be
derived:

$$\text{1st order:} \quad \ln [PM] = -k_{obs} t$$

$$\text{2nd order:} \quad \frac{1}{[PM]} = [PM]_o k_{obs} t + 1$$

where $[PM]_o$ is the total porphyrin concentration
that is the initial metalloporphyrin concentration.

Plots for both orders were sketched for CH_3SO_3H
and it was found that the 1st order plot (Figures
12.1, 12.2, 12.3) gave a far better fit.

$$n\text{th order:} \quad \frac{1}{[PM]} = (n-1) [PM]_o (n-1) k_{obs} t$$
$$+ \frac{1}{[PM]_o (n-2)}$$

The 2nd order plots (Figure 12.1) are concave, signifying that for higher n's, the graph would deviate
more rapidly from a straight line; therefore m must
be equal to 1.

From Figure 12.4, n and k were determined to
be 4 and 1.0 x 10^{-6} sec^{-1} for sulfuric acid and 6.6

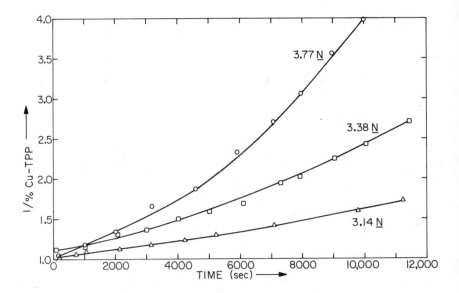

Figure 12.1.   Second order reaction plot for CH_3SO_3H.

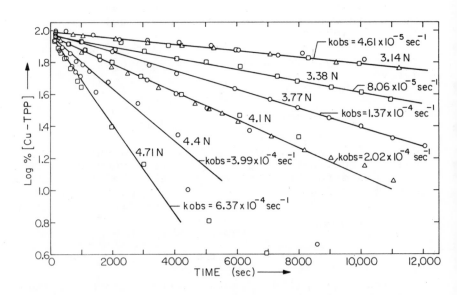

Figure 12.2.   Demetallation kinetics of CuTPP by CH_3SO_3H
at various acid strengths.

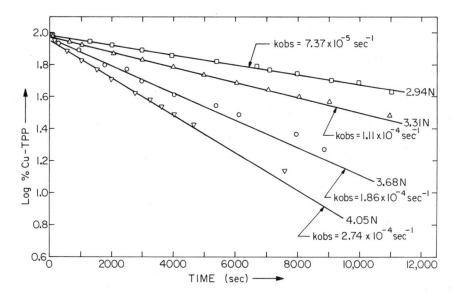

Figure 12.3.   Demetallation kinetics of CuTPP by H_2SO_4.

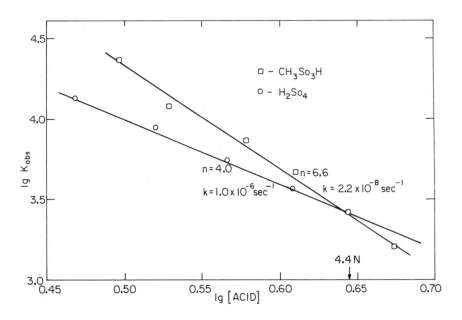

Figure 12.4. Plot of log K_{obs} vs. log acid to determine acid order and rate constant.

and 2.2×10^{-8} sec^{-1} for methanesulfonic acid, respectively. The two graphs intersect at around 4.4 N, signifying that at concentrations below that sulfuric acid is a more efficient demetallation reagent than methanesulfonic acid. The reverse is true for concentrations above 4.4 N.

ACKNOWLEDGMENT

The authors acknowledge Mr. Avak Minnasian for preparation of the copper porphyrins used in this experiment. Partial support from PRF Grant No. 6272-AC2 and JPL-Cal Tech subcontract under NASA NAS 7-100 is also acknowledged.

REFERENCES

1. Yen, T. F. in *Chemistry in Space Research*, R. F. Landel and A. Rembaum, Eds.   (New York: Elsevier Pub. Co., 1972) pp. 105-153.
2. Hayes, J. M. *Geochim. Cosmochim. Acta, 31,* 1395 (1967).
3. Johnson, F. H.   Pacific Conf. on Chemistry and Spectroscopy, Anaheim, 1971.
4. Rho, J. H., A. J. Bauman, T. F. Yen and J. Bonner. *Science 167,* 754 (1970).
5. Rho, J. H., A. J. Bauman, T. F. Yen and J. Bonner in *Proceedings of the Second Lunar Science Conference,* A. A. Levenson, Ed. 3, Vol. 2 (Cambridge, Mass.: MIT Press, 1971) pp. 1875-1878.
6. Baker, E. W.   *J. Am. Chem. Soc., 88,* 2311 (1966).
7. Baker, E. W., T. F. Yen, J. P. Dickie, R. E. Rhodes and L. F. Clark.   *J. Am. Chem. Soc., 89,* 3631 (1967).
8. Vaughan, G. B., E. C. Tynan and T. F. Yen.   *Chem. Geol. 6,* 203 (1970).
9. Yen, T. F. and S. R. Silverman. *ACS Div. Petrol. Chem., Preprints, 14(3),* E32-E39 (1969).
10. Yen, T. F., L. J. Boucher, J. P. Dickie, E. C. Tynan and G. B. Vaughan.   *J. Inst. Petrol., 55,* 87 (1969).
11. Ball, J. S. *et al. ACS Div. Petrol. Chem., Preprints, 1(1),* 241 (1956).
12. Rho, J. H., A. J. Bauman, T. F. Yen and H. Boelteger. Space Biol. Conf., Houston, 1971.
13. Yen, T. F. in *Role of Trace Metals in Petroleum.*   (Ann Arbor, Michigan: Ann Arbor Science Publishers, 1975).
14. Caughey, W. S. and A. H. Corwin.   *J. Am. Chem. Soc., 77,* 1509 (1955).

CHAPTER 13

AMERICAN OIL REFINERIES:
CAN THEY YIELD VANADIUM?

R. P. Fischer
U.S. Geological Survey
Denver Federal Center
Denver, Colorado

This paper presents an idea that might not have occurred to some people in the petroleum industry, namely, the possibility of marketing significant amounts of vanadium recovered from the refining of oil. Contemplation of any commercial venture of this sort requires consideration of the supply-and-demand situation, the technology of recovery, and the economics of recovery. As the vanadium commodity geologist for the Geological Survey, I can attest that the supply of vanadium from domestic producing sources is expected to be short relative to the future expected domestic demand. The problems of the technology and economics of recovery of vanadium from oil, however, are outside my field of competence. Nevertheless, on the basis of a few minor vanadium-recovery operations reported in literature, it would seem reasonable to think that the principles of vanadium recovery are available and that the economics of recovery might be favorable, especially if vanadium is removable with constituents that must be removed to avoid pollution.

VANADIUM SUPPLY AND DEMAND

Currently the bulk of the world's vanadium is recovered as a by-product from deposits that are relatively low in vanadium; resources in deposits of this kind are large.[1] Deposits that can be mined profitably for vanadium alone must contain at least 1% V_2O_5, but not many deposits of this kind are known.

Most of the foreign supplies of vanadium are recovered as a by-product from iron ores. As large supplies of these ores are available, foreign vanadium

201

production probably will continue at a substantial level and the foreign market will be highly competitive.

Figure 13.1 is a generalized graph showing vanadium production from the principal sources for every tenth year from 1910 to 1970. Its purpose is to show the large increase in production, in response to demand, since 1950.

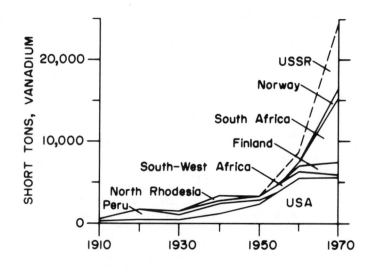

Figure 13.1.   Generalized graph showing the principal sources
of vanadium, 1910-70. Data from published
figures of U.S. Geological Survey, 1910 and
1920, and U.S. Bureau of Mines, 1930-7, except
U.S.S.R. data for 1960 and 1970, which are
estimates by R. P. Fischer.

Domestic supplies of vanadium are obtained from a deposit in Arkansas that is mined for vanadium alone, from some deposits in the western states that yield coproduct uranium and vanadium, and from slags derived from making elemental phosphorus from phosphate rock mined in Idaho. The vanadium-production potential of these deposits does not appear to be adequate to satisfy long-range domestic requirements.

Figure 13.2 shows domestic vanadium production and consumption, 1960 to 1968, and the domestic maximum and minimum forecast demand for vanadium from 1968 to the year 2000, as estimated by Griffith.[2] Cumulative requirements at the maximum demand rate from 1968 to 2000 are 520,000 short tons of vanadium, and 420,000 tons at the minimum rate. Griffith[2] also estimated cumulative production from currently producing domestic sources from 1968 to 2000 to be 115,000 tons of vanadium; the dotted line on Figure 13.2 is my estimate of the yearly rate of this production. If the cumulative figures are correct, they show a domestic vanadium deficiency of some 300,000 to 400,000 tons between 1968 and 2000. If we do not find new domestic sources of vanadium to make up this deficiency, we will have to buy and import this vanadium. At the current price of $1.50 per pound V_2O_5 (about $5,400 per ton vanadium), this deficient vanadium would be worth at least $1.5 billion.

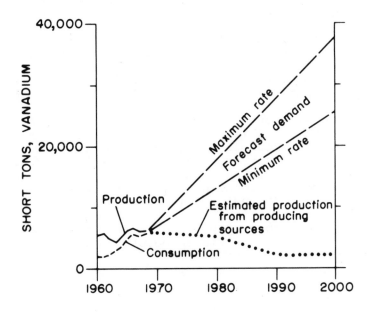

Figure 13.2. Graph showing domestic vanadium production and consumption, 1960-68, and domestic forecast demand and estimated production from currently producing sources, 1968-2000.

VANADIUM IN PETROLEUM AND ITS POTENTIAL

Many crude oils contain little or no vanadium, but some contain as much as several hundred ppm (parts per million).[3,4] This vanadium tends to

accumulate in the residues resulting from the natural
or industrial distillation or combustion of these
oils. Many asphaltites contain about 1% vanadium.[5]
The ash and soot from some oil-burning furnaces have
long been a minor commercial source of vanadium.
Since 1965, the refinery of Canadian Petrofina, Ltd.,
Montreal, Canada, has recovered vanadium commercially
from ash obtained by burning residual oil ("fluid
coke");[6] the crude oil used contains only about
130 ppm V. The recovery of vanadium from the fly
ash of burning a similar residual oil obtained from
refining Athabasca tar sands, Alberta, Canada, has
been proposed;[7] the extracted oil contains about 200
ppm V.[8]

Vanadium is recovered, or its recovery is proposed, by similar practices in the U.S.S.R. and
Hungary.[9,10] Perhaps some vanadium could be recovered
profitably by similar practices in some refineries in
the United States; antipollution controls needed in
the future might enhance the recovery of the vanadium-
bearing ash.

The total amount of vanadium in residual fuel
oil is probably greater than in any other product of
petroleum refining. About 668 million barrels of
fuel oil were consumed in the United States in 1968,
and this material is estimated to have contained
nearly 19,000 tons of vanadium.[11] At the current
price, this amount of vanadium is worth about $100
million. Desulfurization of fuel oil can reduce the
vanadium content about proportionate to the reduction
of sulfur.[12] If fuel oil is desulfurized to reduce
atmospheric pollution, can a significant amount of
vanadium be recovered profitably from the catalysts
used in the desulfurizing process or at some other
stage in this process?

CONCLUSIONS

The need for new domestic sources of vanadium
to meet future demands warrants consideration of
recovering vanadium from petroleum products. The
need to reduce other pollutants in petroleum products
might enhance the technology and economics of vanadium recovery.

REFERENCES

1. Fischer, R. P. in *United States Mineral Resources*, U.S. Geol. Surv., Prof. Paper 820 (1973), p. 679-688.
2. Griffith, R. F. *Mineral Facts and Problems*, U.S. Bureau of Mines, Bull. 650 (1970), p. 417.
3. Whisman, M. L. and F. O. Cotton. *Oil Gas J.*, *69 (52)*, 111 (1971).
4. Nelson, W. L. *Oil Gas J.*, *70 (32)*, 48 (1972).
5. Abraham, H. in *Historical Review and Natural Raw Materials*, Vol. 1, 6th ed. (Princeton, N.J.: Van Nostrand Co., 1960), p. 370.
6. Whigham, W. *Chem. Eng.*, *72 (5)*, 64 (1965).
7. "Athabasca Crude to Hit Pipeline Soon," *Oil Gas J.*, *65 (41)*, 76 (1967).
8. Scott, J., G. A. Collins and G. W. Hodgson. *Can. Inst. Mining Met.*, *Trans.*, *47*, 36 (1954).
·9. Volkova, P. I., N. A. Vatolin, A. A. Ryzhov and N. N. Belyaeva. *Tr. Inst. Met. Sverdlovsk*, *17*, 118 (1969); *Chem. Abs.*, *73 (4)*, abs. no. 17611e (1970).
10. Lovasi, J., M. Miskei, L. Tomcsanyi, P. Siklos and L. Farkas. Hungarian patent, Teljes 1113 (Cl. C. Olg), October 24, 1970; *Chem. Abs.*, *74 (20)*, abs. no. 101219m (1971).
11. Davis, W. E., and Associates. "National Inventory of Sources and Emissions. Arsenic, Beryllium, Manganese, Mercury and Vanadium," Report for Environmental Protection Agency, Contract No. CPA 70-128, 1971, p. 47.
12. Radford, H. D. and R. G. Rigg. *Hydrocarbon Process*, *49 (11)*, 187 (1970).

CHAPTER 14

FREE WORLD SUPPLY AND DEMAND FOR VANADIUM
FROM 1973 THROUGH 1980

F. J. Shortsleeve
Mining and Metals Division, Union Carbide Corporation
270 Park Avenue, New York, New York 10017

Union Carbide has recently completed a study of
the Free World supply and demand for vanadium.[1] The
conclusions of the study are substantially different
from those conducted earlier due to changing market
demand and because of recently announced plans by
the basic producers of the Free World to expand production capacities. This paper presents the results
of the recent study.

Consumption of vanadium has increased steadily
since the early 1960s due principally to the increased demand for vanadium-bearing high strength,
low alloy (HSLA) steels; furthermore, it is expected
that the construction of thousands of miles of large
diameter vanadium steel pipelines, along with increasing use of high strength structural steels, will
continue to improve the demand for vanadium. Consequently, Free World demand is expected to increase
from 38 million pounds of vanadium in 1973 to 51.7
million pounds in 1980.

On the other hand, throughout that period Free
World milling capacity is expected to be well in
excess of Free World demand. The milling capacity
of the present Free World basic vanadium producers
is estimated as 42.5 million pounds* for 1973 and
is expected to increase to 58.8 million pounds in

---

*Vanadium values are the calculated quantities of vanadium
recoverable from vanadium-bearing slags and from V205,
assuming 85% recovery of V205 from slag and 92% recovery in
conversion of V205 to FeV type products.

207

1980. New producers have announced plans to install 13.1 million pounds of additional capacity, bringing the estimated 1980 production to 71.9 million pounds of vanadium, representing 139% of projected demand. Within the United States milling capacity is estimated at 19 million pounds of vanadium. A new producer has announced plans to enter the business in 1976, bringing domestic production capacity to 21.1 million pounds. These figures are well in excess of the 1973 estimated demand of 12.5 million pounds for the United States and Canada, and the production is adequate to satisfy the projected demand of 16.1 million pounds in 1980. In addition to basic vanadium oxide producers, a major domestic converter of ferrovanadium is now producing millions of pounds of vanadium per year for sale in the United States from imported, duty free, steel slags.

The need to reduce pollutants in petroleum products may provide still another source of vanadium; however, in view of present and projected milling capacities of basic producers, the relative economics of recovery of vanadium from petroleum products should be carefully assessed before capital investments are seriously considered.

SOURCES OF VANADIUM

Vanadium is found widely in nature and is more plentiful than copper, zinc or lead. The known, commercially viable ore deposits are very extensive compared with present and projected world demand. The major sources are:

1. vanadium-uranium coproduct carnotite ores of the Colorado Plateau
2. vanadiferous clays in Arkansas
3. titaniferous vanadium-bearing magnetites in South Africa, Russia, Norway and Finland.

Other sources of vanadium are the vanadium-bearing phosphate ores of Idaho, by-product slags of lead mining operations in South West Africa, and bauxite operations in France. Significant quantities of vanadium are also available in boiler scale and other residues resulting from the combustion of vanadium-bearing fuels for power generation.

Vanadium is usually produced as a by-product or coproduct along with uranium, phosphorus, iron, titanium and, in one instance, copper, lead and zinc.

Only the Arkansas vanadiferous clays and a portion
of the titaniferous magnetites of the Bushveld complex
of South Africa are worked for vanadium values only.
At present, magnetites of South Africa, Russia, Norway and Finland account for a substantial portion
of the world vanadium supply, yielding slags containing 15% to 27% V_2O_5. The reserves of present working
deposits contain sufficient vanadium to supply world
requirements many times over for decades to come.

Very extensive, nonproducing, vanadium-bearing
magnetite deposits exist in Australia, New Zealand,
Canada and India similar to those now being worked.
The deposits in the Singhbhum area, Bihar and
Orissa, India are reported to have 20 million tons
of ore reserves averaging perhaps 2% V_2O_5,[2] a quantity sufficient to supply the total requirements of
the Free World for many years.

Slags, rich in vanadium oxide, from magnetite
operations are transported to major industrial
areas of the Free World to be processed to vanadium
oxide, which is converted to ferrovanadium by
aluminothermic or electric furnace reduction. In
some instances, the slags are reduced directly to
iron-vanadium steelmaking additives by duplex electric furnace processing. Ferrovanadium products
containing 25% to 85% vanadium are used for steelmaking throughout the world, and the impurity elements,
Al, Si, Mn, P, S and O, are maintained at required
levels. In the United States a product made by carbon
reduction of pure vanadium oxide in vacuum is used
extensively. It contains about 84% vanadium and 12%
carbon.

FREE WORLD MARKETS

The steel industry accounts for over 90% of the
present Free World demand for vanadium, and increasing demand in steel applications is responsible for
most of the projected growth. Vanadium is used also
as an alloying element in titanium-base alloys and as
vanadium catalysts by the chemical industry. Table
14.1 displays estimated vanadium usage in 1973 by
major end use for the Free World.

In steel applications, vanadium is used in the
production of HSLA steels, tool and die steels, open
die forgings and to some extent as a strengthening
agent in plain carbon steels. The titanium industry
employs vanadium as an alloy addition to titanium-
base alloys used principally in air-frame and gas

Table 14.1

Estimated 1973 Free World
Vanadium Consumption by Major End Uses

End Use	Consumption lb., V 10^6	Total Free World Total %
Steel		
Alloy steel	5.6	16
Tool steel	9.0	26
Carbon steel (rebar, piling)	2.2	6
HSLA (incl. line pipe), castings, forgings	15.8	45
Total steel uses	32.6	93
Other		
Titanium, chemicals	2.3	7
Total-all uses	34.9	100
Exports to Communist nations	3.1	
Consumption	38.0	

turbine engine construction, and usage in the United States approaches one million pounds of vanadium per year. In the chemical industry, vanadium is used chiefly as a catalyst in the production of sulfuric acid, maleic anhydride, phthalic anhydride and adipic acid.

FREE WORLD CONSUMPTION

Table 14.2 shows estimated 1973 and forecast 1980 vanadium consumption in Free World markets. Total Free World demand is 34.9 million pounds of vanadium in 1973 and is projected to increase at 4.6% per year to 47.8 million pounds in 1980. Exports to Communist countries are estimated at 3.1 million pounds in 1973 and 3.9 million pounds in 1980. The total Free World market, therefore, is 38.0 million pounds in 1973 and is expected to grow to 51.7 million pounds in 1980.

Table 14.2

Vanadium Consumption by Major Free World Markets
(In Millions of Lbs. Vanadium)

Major Market Areas	V Consumption 1973	1980	Average Annual Growth Rate %
Canada and U.S.A.	12.5	16.1	3.7
Western Europe	14.9	19.9	4.2
Other Free World nations	7.5	11.8	6.3
Total Free World	34.9	47.8	4.6
Export to Communist nations	3.1	3.9	3.3
Total Free World Market	38.0	51.7	4.5

The consumption of vanadium in Western Europe is forecast to increase at a rate of 4.2% per year from 14.9 million pounds in 1973 to 19.9 million pounds in 1980. The growth in demand is due principally to the anticipated use of vanadium steel for the construction of gas and petroleum pipelines and the emergence of strong national steel industries in certain countries of Western Europe. Furthermore, consumption of vanadium-bearing tool and die steels is expected to increase the demand for vanadium by the specialized steel industries of Austria and Sweden.

Demand for vanadium in the United States and Canada is nearly equal to that of Western Europe. The steel industry in the United States has not yet installed the equipment necessary to fabricate the new large diameter (over 42" O.D.) line pipe being specified for many major line pipe projects. On the other hand, Canada, Japan, West Germany and Italy are well equipped to produce the pipe and are expected to be the major recipients of line pipe orders. As a result, increasing demand for vanadium in the United States is expected to be modest and, in the absence of other major developments, such as automotive HSLA use, will remain at 3% per year through 1980. The potential application of HSLA steels to automobile bumpers, frames, engine mounts, etc., could increase United States vanadium consumption by 10-15% or 1 to 1.5 million pounds per year. However, competition from other materials precludes an accurate appraisal at this time.

FREE WORLD PRODUCTION

The Free World production capability of present producers is expected to increase from 42.5 million pounds of recoverable vanadium in 1973 to 58.8 million pounds in 1980. The projection includes the expansion of Union Carbide's South African facility to a level of 6.2 million pounds of recoverable vanadium in 1976 and 10.3 million pounds in 1978. Furthermore, the production capability of new basic producers is estimated as 13.1 million pounds in 1980, bringing total Free World production capacity to 71.9 million pounds at that time.

Several vanadium properties are being evaluated as additional sources of vanadium, but firm determinations have yet to be made relative to economic viability. Substantial vanadium reserves are indicated for many of these properties, and a decision to put any one into production would result in a significant addition to the projected capacity.

SUPPLY VERSUS DEMAND

The projected Free World supply and demand are compared in Figure 14.1 and in Table 14.3. Production capacity is estimated at 42.5 million pounds of vanadium in 1973 and consumption at 38 million pounds, a surplus of 4.5 million pounds or 12%. In fact, 1973 production capability is sufficient to supply Free World needs through 1975. The projected mill capacity of the present producers is adequate to satisfy the expanding demand for vanadium through 1980, and the additional production anticipated by new producers is likely to result in a supply well in excess of Free World demand.

Mill capacity within the United States is estimated at 19.0 million pounds of vanadium for 1973 through 1980. The United States capacity is well in excess of the 1973 demand for 12.5 million pounds for the United States and Canada and is adequate to satisfy the projected demand of 16.1 million pounds in 1980. Furthermore, a new producer has announced plans to produce vanadium oxide in the United States, bringing the total production capacity to 21.1 million pounds of vanadium by 1976. In addition to the basic United States V205 production capability, a major domestic converter is now producing millions of pounds of ferrovanadium per year from duty free steel slags imported from South Africa, a source, that if further developed, could satisfy the entire projected domestic demand for vanadium.

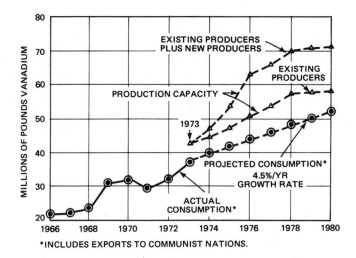

Figure 14.1.   Vanadium supply/demand--Free World 1973-1980.

Table 14.3

Free World Supply and Demand
(In Millions of Pounds of V)

	Free World Production Capacity	Free World Demand	Supply as a % of Demand
1973	42.5	38.0	112
1974	46.0	39.7	116
1975	52.9	41.5	127
1976	62.4	43.4	144
1977	66.2	45.3	146
1978	70.9	47.4	150
1979	71.9	49.5	145
1980	71.9	51.7	139

CONCLUSIONS

In the foregoing, it may be seen that the milling capacity of the domestic facilities and, indeed, throughout the Free World will be more than adequate to satisfy the projected demand for vanadium. The recovery of vanadium from petroleum products may represent still another source of vanadium that may or may not be suitable for processing through existing mills. Under the circumstances, the relative economics of recovery of vanadium from petroleum products should be carefully assessed before capital investments are seriously considered.

REFERENCES

1. Brown, C. M., L. C. Johnston and G. J. Goetz. "Vanadium—Present and Future," Union Carbide Corporation, Mining and Metals Division, presented before the Canadian Institute of Mining and Metallurgy on August 28, 1973.
2. Fischer, R. P. "Vanadium," U.S. Geol. Survey Prof. Paper 820.

INDEX

INDEX

acetic acid 197
acids 195
acid treatment 27
allomerization 17,18
analytical method selection 60
  (see also trace analysis)
anisotropy-isotropy 172
antimony 38,40,43,48,92,113
  as a lubricant additive 141
  content in asphalt 140
  content in crude oils 136
  physiological effects 126
  (see stibines)
antioxidant additives 141
  (see also lubricant
  additives)
antiwear agents 141
arsenic 31,38,48,92,113,114
  content in crude oils 136
  physiological effects 126
Ascidia mentula 19
asphalt 138-140
  concrete 140
asphaltenes 7,13,22,33,35,40,
    41,43,44,45,48,54,
    55,99,111,120
Athabasca tar sands 164,165,
    204
atmospheric tower bottoms 161
  (see also molybdenum)
atomic absorption
  analysis 154,156-159,161
  background effects 157
  flame photometry 79,81
  spectrophotometry 163
  (see also instrumental
  methods)

atomic emission
  flame photometry 79,81,82
  spectroscopy 77
  (see also instrumental
  methods)

Bachaquero crude 174
barium
  as a lubricant additive 141
  content in crude oils 136
  physiological effects 126
bauxite 208
Baxterville resin 12
bitumens 170
boron
  as a lubricant additive 141
  physiological effects 127
Boscan crude 38,151,155,174,
    187
bromine 113,114

cadmium 93,101,135
  as a lubricant additive 141
  content in crude oils 136
  physiological effects 127
californium 68
carboniferous era 19
cation-radicals 188
cesium 92,113
chlorine 10,113,114
chlorophyll 10,13
chromium 31,38,92,113,118
  as a lubricant additive 141
  content in asphalt 140
  content in crude oils 137
  physiological effects
    127

217

DATE LOANED

APR 4 1978			